Jonas Johannes Hüther

The Impact of Recycling on the Fibre and the Composite Properties of Carbon Fibre Reinforced Plastics

Schriftenreihe
des Instituts für Angewandte Materialien
Band 84

Karlsruher Institut für Technologie (KIT)
Institut für Angewandte Materialien (IAM)

Eine Übersicht aller bisher in dieser Schriftenreihe erschienenen Bände finden Sie am Ende des Buches.

The Impact of Recycling on the Fibre and the Composite Properties of Carbon Fibre Reinforced Plastics

by
Jonas Johannes Hüther

KIT Scientific Publishing

Karlsruher Institut für Technologie
Institut für Angewandte Materialien

The Impact of Recycling on the Fibre and the Composite Properties of Carbon Fibre Reinforced Plastics

Zur Erlangung des akademischen Grades eines Doktor-Ingenieurs von der KIT-Fakultät für Maschinenbau des Karlsruher Instituts für Technologie (KIT) genehmigte Dissertation

von M.Sc. Jonas Johannes Hüther

Tag der mündlichen Prüfung: 6. September 2019
Referent: Prof. Dr.- Ing. Kay André Weidenmann
Korreferenten: Univ.- Prof. Dipl.- Chem. Dr. Alexander Bismarck
Prof. Dr.- Ing. Peter Elsner

Impressum

KIT Scientific Publishing

Karlsruher Institut für Technologie (KIT)
KIT Scientific Publishing
Straße am Forum 2
D-76131 Karlsruhe

KIT Scientific Publishing is a registered trademark of Karlsruhe Institute of Technology.
Reprint using the book cover is not allowed.

www.ksp.kit.edu

Print on Demand 2020 – Gedruckt auf FSC-zertifiziertem Papier

ISSN 2192-9963
ISBN 978-3-7315-0983-7
DOI 10.5445/KSP/1000098911

The Impact of Recycling on the Fibre and the Composite Properties of Carbon Fibre Reinforced Plastics

Zur Erlangung des akademischen Grades

Doktor der Ingenieurwissenschaften

der Fakultät für Maschinenbau

Karlsruher Institut für Technologie (KIT)

genehmigte

Dissertation

von

Jonas Johannes Hüther (M.Sc.)

Tag der mündlichen Prüfung: 06. September 2019

Referent: Prof. Dr.- Ing. Kay André Weidenmann

Korreferent: Univ.- Prof. Dipl.- Chem. Dr. Alexander Bismarck

Korreferent: Prof. Dr.- Ing. Peter Elsner

Dedicated to Georgia & Manfred

Artikel 5, Absatz 3:

> Kunst und Wissenschaft, Forschung und Lehre sind frei. <

In recognition of the 70th anniversary of the
Basic Law for the Federal Republic of Germany

Kurzfassung

Die vorliegende Arbeit beschäftigt sich mit der Methodenentwicklung im Bereich der Einzelfaserprüfung technischer Fasern sowie dem Einfluss des pyrolytischen Recyclings auf Verbundwerkstoffe. Der Recyclingeinfluss wurde auf Einzelfaserebene ebenso untersucht wie auf Verbundwerkstoffebene, wobei Verbundwerkstoffe auf Basis neuer und recycelter Fasern im Bulk Moulding Compound Prozess hergestellt wurden und Gegenstand der Materialcharakterisierung waren. Weiterhin wurde die Faser-Matrix-Grenzfläche und die Faseroberfläche untersucht. Dadurch konnten alle Skalen von der Einzelfaser- bis zur Verbundebene betrachtet werden.
Zur Verbesserung der Einzelfaserprüfung von Kohlenstofffasern wurden der experimentelle Prüfaufbau und die Testmethoden weiterentwickelt. Die Einzelfasern lassen sich hiermit in variablen Einspannlängen untersuchen. Anstelle von Kreisflächen lassen sich mit dem überarbeiteten Konzept elliptische Querschnitte annähern, was der wahren Form von Kohlenstofffasern wesentlich näher kommt. Die Messung der Dehnung von Kohlenstofffasern konnte mittels digitaler Bildkorrelation direkt an der Faser durchgeführt werden, was Messfehler durch die Maschinensteifigkeit ausschließt. In der Folge war es durch die entwickelte Methode möglich, den E-Modul und die Querschnittsfläche von Kohlenstofffaser besser zu korrelieren, das nicht-Hook'sche Verhalten klar zu erkennen und eine Messung der Querkontraktion zu ermöglichen
Neben der Methodenentwicklung stand der Einfluss des Recyclings von kohlenstofffaserverstärkten Kunststoffen im Fokus. In der Wertschöpfungskette der kohlenstofffaserverstärkten Kunststoffe fallen sowohl trockener Produktionsabfall als auch vollständig imprägnierte Bauteile an. Im Rahmen der Arbeit

gelang es, beide Abfallklassen im gleichen Herstellungsprozess, dem direkten Bulk Moulding Compound Prozess, zu verarbeiten. Sowohl aus Verschnittresten als auch aus End-of-Life-Bauteilen wurden Platten mit Wirrfaserverteilung und einem Fasergewichtsanteil von 55 % hergestellt, die zum Beispiel im Interieur von Fahrzeugen zum Einsatz kommen könnten. Die mechanischen Kennwerte unterlagen dabei materialübergreifend starken Schwankungen mit Steifigkeiten zwischen 15 und 30 GPa sowie Festigkeiten zwischen 30 und 120 MPa.

Die Pyrolyse wurde beim konventionellen Anbieter *carboNXT GmbH* durchgeführt, um anwendungsnahe Einflüsse abzubilden. Ein Einfluss der Pyrolyse zeigte sich in erster Linie auf Faserebene: Während der E-Modul der Einzelfasern im Wesentlichen unverändert blieb, sank die Festigkeit um etwa 20% ab. Die recycelten Fasern zeigten sowohl eine höhere nanoskalige Rauhigkeit als auch eine chemisch aktivere Oberfläche. Die scheinbare Grenzflächenscherfestigkeit zwischen Faser und Matrix war für beide Fasertypen ähnlich, aufgrund der erhöhten Rauheit kam es bei der Verwendung recycelter Fasern jedoch zu höheren Reibwerten zwischen Faser und Matrix.

Auf Verbundebene konnte bezüglich Steifigkeit und Festigkeit kein Unterschied zwischen trockenen Verschnittfasern und recycelten Fasern festgestellt werden. Die Streuung der mechanischen Kennwerte der Materialien resultiert aus querliegenden Fasern und harzreichen Regionen, die als Schwachstellen das Versagen dominieren. Dieses Versagensverhalten erschwert zwar die Nutzung der BMC Bauteile in lasttragenden Strukturen, erlaubt aber eine effiziente Nutzung aller kohlenstofffaserhaltigen Recyclingabfälle in einem einzigen, kostengünstigen Prozess. Hiermit ergibt sich die Möglichkeit, dem gesellschaftlich-technischem Ziel, im Bereich des Automobilbaus keinerlei Abfälle mehr zu deponieren, Rechnung zu tragen.

Abstract

The work at hand deals with the development of advanced procedures in the field of single fibre testing as well as the influence of carbon fibre recycling on the resulting properties of the fibres and bulk moulding composites produced from them. Different scales of composites are characterised: the fibres and their surfaces, the fibre/matrix-interface and the composite as a whole. In order to extend single fibre testing, the experimental devices and the testing procedure have been enhanced in some aspects: the single fibres can be investigated with adaptable gauge lengths. Instead of assuming circular fibre cross-sectional areas, the modified procedure applies elliptical approximations to get closer to the actual shape of carbon fibres. A technique to study fibre strain directly on the fibre by means of digital image correlation is presented, reducing inaccuracies caused by the system compliance. As a result, it was possible to correlate the cross-sectional area and the Young's modulus of carbon fibres. The non-Hookean behaviour of carbon fibres and their Poisson's ratio could thus be determined.

Along the value chain of carbon fibre composite production, waste is generated in the form of non-impregnated (dry) fibre waste, pre-impregnated parts and also end-of-life structures. Within the frame of the work at hand, it was feasible to recycle dry fibre waste and end-of-life waste in the very same production process, namely the bulk moulding compound process. Dry fibre waste and recycled fibres were used to manufacture panels with randomly orientated fibres and a fibre weight content as high as 55 %. Regardless of the fibre-matrix-combination, the mechanical properties showed large variations, ranging from 15 to 30 GPa in terms of stiffness and 30 to 120 MPa in terms

of tensile strength. Nonetheless, the materials qualify for use in the interior of vehicles, increasing the sustainability in this sector.
The fibres were reclaimed by means of conventional pyrolysis by the company *carboNXT GmbH* to obtain application-relevant results. An impact of recycling was mainly found on fibre level: while the Young's modulus remained unchanged, the tensile strength of recycled carbon fibres was reduced by approximately 20 % compared to the reference. In addition, the nano-scale roughness and the chemically active surface were increased. The interface between fibre and matrix was similar in terms of shear strength for recycled and virgin fibres. Nonetheless, for recycled fibres, friction-related properties were increased. Between virgin and recycled fibres, no significant difference in the tensile properties of the composites was found. The results showed high scatter as a consequence of the non-uniform structure of bulk moulding compounds (BMC). Fibres are apparent as bundles in BMC, with misaligned fibre bundles and resin-rich regions dominating the failure behaviour. Both are independent of the state of the fibre but an intrinsic property of moulding compounds.
On the one hand, this failure mechanism complicates the usage of BMC materials as load-carrying structures. On the other hand, the BMC route is an efficient and straightforward method for recycling of dry waste and end-of-life carbon fibres in one and the same process. Thus, the strategy evaluated in the work at hand allows the automotive industry to recycle and reuse carbon fibres by producing secondary components for vehicles instead of disposing them in landfills.

Acknowledgement

For me, it had been three and a half memorable and remarkably pleasant years as a PhD candidate at Karlsruhe Institute of Technology. During this time, there have been dozens of people who offered their support, encouragement and inspiration and I want to thank all of you.

First of all, I would like to thank my reviewers:
Prof. Dr.- Ing. Kay André Weidenmann, for all his ideas and support, from composite production to material characterisation. I am also very thankful for your corrections and advices for our papers and my thesis. I highly appreciate your support in times I did not get ahead.
Univ.- Prof. Dipl.- Chem. Dr. Bismarck for three inspiring weeks at the University of Vienna, giving me an additional perspective on my materials and experiments. Thank you very much for being an assessor of my work, providing me with numerous ideas on how to optimise my thesis.
Prof. Dr.- Ing. Elsner for evaluating my thesis and the successful and entertaining collaboration in our teaching.

I would like to thank all my current and former colleagues at IAM-WK for numerous discussions, inexhaustible support and our collective brain-storming for new ideas. Please understand that I cannot list you all name by name but still, I would like to express my greatest gratitude! I am very glad to have you as colleagues and friends. In particular, I would like to thank all co-authors of my scientific publications and my project partners at FAST and Fraunhofer ICT.

I am very grateful that I could work with many students within the past few years. I would love to thank all the research assistants and all the students who worked on their thesis within my project. To name all of you in chronological order: Ina Kohlschreiber, Julian Hüttl, Yasmine Kechaou, Theresa Jenne, Magdalena Lenz, Lukas Mayer, Wendelin Buderer, Markus Muth, Philipp Schumann, Daniel Esse, Alexander Klauber, Balduin Katzer, Daniel Mayer, Florian Schwarz and Nikol Vlavaki. Thank you for your help and support!

I would highly like to express my sincere appreciation to Dr. Andreas Reeb, Dr. Wilfried Liebig, Prof. Dr. Povl Brønsted and Dr. Klaudia Lichtenberg who were always there when I needed advice and who supported me on the long road to become the scientist I am now.

I highly cherish the help by research partners:
Johanne Hesselbach at the Faserinstitut Bremen for our discussion on fibre properties and especially for running tensile tests and pull-out tests with my fibres.
Dr. Moosburger-Will, Eva Laukmanis and Michael Schulz, from the University of Augsburg, for AFM investigation and push-out tests on recycled and virgin materials.
Sven Caspart for our productive debates concerning the mathematics behind the cross-sectional areas of fibres and Siegfried Galkin for discussing their evaluation.
Apl. Prof. Dr. habil. Reinhard Schneider, Dr. Erich Müller and Dipl.Ing. Volker Zibat for their support with SEM and TEM experiments at the Laboratory for Electron Microscopy, KIT.
I highly appreciate the support by André Grube and Thomas Graebe, both employees at GOM, in assisting me with questions related to digital image correlation and for the scientific discussion how to apply the technique on fibre level.

Univ.-Prof. Dipl.-Chem. Dr. Bismarck, Dr. Andreas Mautner and their team at the University of Vienna for all their support during and after my exchange to Vienna.
Special thanks to my office co-worker Markus Muth, for all his support and patience in the month before the submission of my thesis, and for keeping me from going insane. Many thanks to Rebeccah and Georg Löwisch for taking a closer look at grammar and language. I would also like to thank Julius Faubel for shooting a short movie about the SiFiT device and Paul-Leo Lochner for many scientific discussions.

Many of those acknowledged here are about to become a PhD and I cross my fingers that you all bring your dissertations to a successful completion. I am looking forward to supporting you whenever you need my help.

The funding of my research by Baden-Württemberg Stiftung gGmbH and the Ministry of Finance Baden-Württemberg is highly appreciated.

And finally, last but by no means least, I would like to thank my parents, my sister and my friends for their love and support.

Karlsruhe, July 2019 *Jonas Johannes Hüther*

Abbreviations

ABS	acrylonitrile butadiene styrene
AFM	atomic force microscope / microscopy
ASTM	American Society for Testing and Materials
avg	average; evaluation procedure based on mean values
BET	Brunauer-Emmett-Teller theory
BMC	bulk moulding compound
BS	British standard
BSU	basic structural unit
CAD	computer-aided design
CF	carbon fibre
CFF	carbon fibre filler
CFRP	carbon fibre reinforced plastic / polymer
CNC	computerised numerical control
CSA	cross-sectional area
CT	computed tomography
deb	debatable
DIC	digital image correlation
DMA	dynamic mechanical analysis
DMC	dough moulding compound; equal to BMC
DSM	direct strain measurement
DVS	dynamic vapour sorption

E GF	electrical glass fibre (standard glass fibre)
EDX	energy-dispersive X-ray spectroscopy
EP	epoxy resin
FT-IR	Fourier transform infrared spectroscopy
GOM	Gesellschaft für optische Messtechnik mbH
HCl	hydrogen chloride
HiPerDiF	high performance discontinuous fibre
HM	high modulus (carbon fibre)
HP-RTM	high pressure resin transfer moulding
HRTEM	high resolution transmission electron microscopy
HS	hybrid resin
HT	high tenacity (carbon fibre)
IAM	Institute for Applied Materials (KIT)
iGC	inverse gas chromatography
ICT	Fraunhofer Institute for Chemical Technology
IM	intermediate modulus (carbon fibre)
ISM	indirect strain measurement
ISO	International Organization for Standardization
IUPAC	International Union of Pure and Applied Chemistry
KCl	potassium chloride
KIT	Karlsruhe Institute of Technology
KOH	potassium hydroxide
LDS	laser diffraction sensor
LEM	Laboratory for Electron Microscopy (KIT)
LSM	laser scanning micrometer
n.d.	not declared

OM	optimal microscope / microscopy
PA	polyamide
PAN	polyacrylonitrile
PET	polyethylene terephthalate
PP	polypropylene
PUR	polyurethane
rCF	recycled carbon fibre
rCFRP	polymer reinforced with recycled carbon fibres
rEP	composite consisting of recycled CF and epoxy resin
rHS	composite consisting of recycled CF and hybrid resin
R-I	fraction of dry fibre waste
R-II	production waste
R-III	end-of-life waste
ROI	region of interest
RTM	resin transfer moulding
SD	standard deviation
S GF	high strength glass fibre
SAED, SAD	selected area diffraction patterns
SEM	scanning electron microscope / microscopy
SiC	silicon carbide, as fibres or particles
SiFiT	single fibre testing (device / procedure)
SIMS	secondary ion mass spectroscopy
SMC	sheet moulding compound
smo	smoothening effect
TEM	transmission electron microscope / microscopy
TGA	thermogravimetric analysis

UHM	ultra high modulus (carbon fibre)
UP	unsaturated polyester
UPPH	unsaturated polyester polyurethane hybrid resin
UV	ultra violet (light)
UVI	University of Vienna
vCF	virgin carbon fibre
vEP	composite consisting of virgin carbon fibres and epoxy resin
vHS	composite consisting of virgin carbon fibres and hybrid resin
wei	evaluation based on Weibull's theory

Symbols

symbol	description	unit
a_L, c_L, $c_L/2$	lattice parameters	nm
a	semi-major axis of an ellipse	µm
a_{ROI}, b_{ROI}	dimensions of the region of interest	µm
A_0	initial cross-sectional area	μm^2
A_{alt}	cross-sectional area, determined by alternative procedure	μm^2
A_C	area share related to highly ordered regions	%
A_{CSA}	cross-sectional area of a circle	μm^2
$A_{ellipse}$	cross-sectional area of an ellipse	μm^2
A_e	embedded fibre area	μm^2
A_{fibre}	cross-sectional area of a carbon fibre	μm^2
A_i	circular cross-sectional area based on inner radius	μm^2
A_o	circular cross-sectional area based on outer radius	μm^2
A_m	cross-sectional area, initial measurement	μm^2
A_{ROI}	area of the region of interest	μm^2
A, B	atomic layers	–
b	semi-minor axis of an ellipse	µm

symbol	description	unit
B1 and B2	regions where beads are applied	–
d	diameter (in general)	µm
d_{fibre}	diameter of a fibre	µm
d'	projection / distance / apparent diameter	µm
$\Delta d'$	incremental apparent diameters	µm
d'_c	corrected diameter	µm
d_{app}	apparent diameter of a regular circle	µm
$d_{app,0}$, $d_{app,20}$	apparent diameters in directions 0° and 20° , respectively	µm
$d_{avg,0}$	average apparent diameter	µm
Δd	variation in diameter	µm
D	region where LDS measurements are run	–
E	Young's modulus (in general)	GPa
E_0 and E_a	resting Young's modulus	GPa
$E_{0,lin}$	apparent Young's modulus obtained by linear regression	GPa
$E_{0,2nd}$	apparent Young's modulus obtained by second-order polynomial regression	GPa
E_ε	Young's modulus at strain ε	GPa
E_{mes}	experimentally measured Young's modulus	GPa
E_{gra}	Young's modulus of graphite	GPa
f	fibre-related correlation factor	–
F	force	N
F_i	failure probability	–
F_{max}	maximum force	N

symbol	description	unit
ΔF	variation of force	N
G_{ic}	critical interfacial energy release rate	$\mathrm{J\,m^{-2}}$
l_e	embedded fibre length	µm
l_{micro}	embedded fibre length (microbond test)	µm
l_{pull}	embedded fibre length (pull-out test)	µm
l_{push}	embedded fibre length (push-out test)	µm
l_0	initial length	mm
Δl	length variation	mm
L	distance between sensor and object	mm
L_s	distance of supporting bars	mm
n	positive integer, here count of individual tests	–
n_s	positive integer, amount of selected fibres	–
m	Weibull modulus	–
M	material index	–
p_z	p-orbital in z-direction	–
P, Q, R, S	geometric points	–
P', P'', X'	geometric points	–
$\overline{PQ}$, $\overline{RS}$	tangents	–
r_e	'radius' of an ellipse	µm
r_{fibre}	radius of a fibre	µm
r_{min}, r_{max}	minimal and maximal radius	µm
r_o, r_i	outer and inner radius	µm
R_a	mean roughness	nm

symbol	description	unit
R_{max}	maximum roughness	nm
R_z	roughness depth	nm
R^2	coefficient of determination	–
s	deflection of central point	mm
sp^2	sp^2-orbital	–
S1, S2	geometric sets	–
t	thickness	mm
t_{dwell}	dwell time	s
Δt	time step	s
T_g	glass transition temperature	°C
U, U', W, W'	carbon atoms in lattice structure of graphite	–
wt.-%	(fibre) weight content	%
x	distance	mm
x', y'	segments	mm
x_{d1}, x_{d2}	distance between adjacent fibres	µm
x_{P1}, x_{P2}	position on circumference of fibre surface	µm
x_1, x_2, x_3	degrees of freedom for SiFiT setup (axial)	mm
ΔZ_n	distance between nth nodes	mm
β	rotation angle	°
γ	rotation angle	°
γ_s^{AB}	specific surface energy	$\mathrm{mJ\,m^{-2}}$
γ_s^D	dispersive surface energy	$\mathrm{mJ\,m^{-2}}$

symbol	description	unit
γ_s^T	total surface energy	$\mathrm{mJ\,m^{-2}}$
$\gamma_{s(max)}^T$	maximum total surface energy	$\mathrm{mJ\,m^{-2}}$
$tan(\delta)$	loss factor	–
ε	strain perpendicular (in general)	–
ε_d	strain perpendicular to fibre direction	–
ε_l	strain in fibre direction	–
$\Delta\varepsilon$	strain variation (in general)	% or –
λ	wavelength	nm
ν	Poisson's ratio	–
ρ	density	$\mathrm{g\,cm^{-3}}$
σ_{pre}	nominal pre-cycle stress	MPa
σ_{ref}	nominal stress on reference level	MPa
σ_1, σ_2 , σ_3	nominal stress at different stress levels	MPa
σ_n	nominal stress at nth stress level	MPa
$\Delta\sigma_T$	nominal stress step	MPa
σ_W	Weibull strength	GPa
τ	interfacial shear stress	MPa
τ_{app}	apparent interfacial shear strength	MPa
τ_d	local interfacial shear strength	MPa
τ_f	interfacial frictional stress	MPa
τ_{IFSS}	interfacial shear stress	MPa
φ, φ'	rotation angle	°
φ_s	rotational step	°
ϕ_1	degree of freedom for SiFiT setup (rotational)	mm
Ω	position of overflow channel	–

Contents

1 Introduction and motivation

1.1 Motivation to recycle fibre reinforced plastics

Humankind is confronted with severe climate changes and scarcity of raw materials. As a consequence, resource efficiency and sustainability have to be encouraged. In this context, reducing the quantity of raw materials deployed is one solution, expanding and improving recycling technologies is another. Reasons and motivation to recycle are manifold: legal regulations have to be met and social and ecological reflections motivate increased amounts of recycling material. Last but not least, supporting recycling will lead to economic benefits as providing primary materials is costly.

While recycling has been extensively used for materials like polymers, aluminium and paper for decades, recycling of reinforced plastics is less developed and therefore less common at the present time. Due to the high complexity of composite recycling, new strategies of recycling and testing methods are being researched. The aim of engineers to design fibre composites with an enduring cohesion of matrix and reinforcement for the service life is contrary to the demands of effortless recycling. In spite of these disadvantages, fibre composites are beneficial in terms of the mileage of vehicles as they help reducing emissions and gasoline consumption and promote eco-friendly travelling. In order to fully grasp the scientific question behind the work at hand, one must be aware that this contradiction makes the recycling of composite materials more difficult.

The work at hand aims at creating a deeper understanding of changes undergone by carbon fibre composite materials due to recycling. It consists of two experimental parts, as two different goals were to be achieved. The first is the enhancement of methods to evaluate the mechanical behaviour of single carbon fibres (Chapter 3). The second is understanding the influence of recycling of carbon fibres and their composites (Chapters 4, 5 and 6).

Concerning recycling strategies, two different materials are investigated: non-impregnated carbon fibre fabric waste,[1] accruing during primary productions and end-of-life waste, where the fibres have been reclaimed by conventional pyrolysis techniques.

Fibre and composites are investigated on different levels, from microscopic to macroscopic scale: for single fibres, cross-sectional areas, tensile properties and surface properties are tested. Different microscopic techniques and chemical analysis are used to gain insight into the structure-property relationship of carbon fibres. The interface within the composite is examined and relevant composite properties are investigated.

From the viewpoint of material science, recycling is an interesting topic. The relationships between process, structure and properties may vary due to the recycling processes and it is always worth taking a look at the impact of recycling on a material.

The carbon fibre primarily used in this work, PANEX 35 or PX35 by *Zoltek Toray Group*, is advertised by the company as being *"the low-cost commercial carbon fibre that has revolutionised the carbon fibre industry and paved the way for carbon fibre to break into a variety of new applications"*[2]. By way of example from the automotive sector, approximately 20 kg of this type of fibre are used in the Swedish electric car project UNITI[3,4]. Applications in automotive industry and the availability of PANEX 35 were among the reasons for choosing this type of fibre for the experimental investigations.

To utilise recycled carbon fibres, two main goals have to be achieved. Firstly,

[1] Hereinafter, non-impregnated fibre waste will also be referred to as *dry fibre waste*.

[2] http://zoltek.com/products/px35 - 07.05.2019; *Original statement in American English*

[3] http://zoltek.com/wp-content/uploads/2018/03/Uniti-Sales-Sheet.pdf - 07.05.2019

[4] https://www.uniti.earth/company/partners/ - 07.05.2019

mechanical properties and performance as close as possible to a comparable material consisting of virgin components need to be guaranteed. Secondly, the lower the cost of the recycled fibres, the more likely they can compete with other materials, for instance composites reinforced with virgin glass and carbon fibres.

1.2 Contribution to the current state of research

This doctoral thesis aims to provide new insights into the structure-property relationships of carbon fibres and reinforced plastics containing recycled carbon fibres and dry fibre waste, produced via the bulk moulding compound process. In detail, it is dedicated to:

- Development of a versatile single fibre tensile testing method and apparatus, including a technique to measure strains directly on the fibre. Principles of full-field digital image correlation are transferred to fibre level (Chapter 3).
- An improved approximation of the real cross-sectional shape of carbon fibres to yield better input for tensile experiments. Enhanced measurements of both, tensile strain and the cross-sectional area, will allow for more accurate determination of the Young's modulus, the consideration of non-Hookean effects, and the proficiency of determining the Poisson's ratio of single fibres. A better knowledge of the elastic properties enables improvement of micromechanical models.
- The relation between fibre properties and the behaviour of thermosets reinforced with randomly-orientated carbon fibres will be evaluated for virgin and recycled materials. This involves single fibre properties, the characteristics of the surface, the fibre/matrix-interface and the properties of the BMC composite (Chapter 6 and 7).

- The development of a direct bulk moulding compound process, being capable of processing dry fibre fabric waste, recycled carbon fibres and carbon fibre filler. This corresponds to a holistic approach to the recovery of all size fractions of fibre waste (Chapter 8).

From these contributions to the current state of the research, the scientific question are derived in Section 2.4.

1.3 Introductory remarks

A few remarks shall be given here to clarify some terms. Between different years of publication and different areas of studies, the word *"filament"* is used for varying objects; sometimes it refers to a single fibre, sometimes to a cluster of several single fibres, for instance in a yarn. Within this work, "filament", if used, will refer to a single fibre. A single fibre is hereby defined as an object that cannot be broken down to thinner units without destroying its original structure. Another misnomer found in literature is the term *fibre diameter*. "Diameter" would only be appropriate for objects having a perfectly circular shape. Whenever possible, *cross-sectional areas* are used instead in the investigations of this work. Here, "diameter" is only used when quoting other works and to state that an *apparent diameter* is considered. In this context, "apparent diameter" refers to the distance between both edges of the 'shadow' that a fibre would cause [1]. Diameter will always refer to the simplification that a perfectly circular fibre shape is assumed. To allow for comparison with the state of research, diameter readings found in literature are translated to circular cross-sectional areas where needed. When discussing literature in the field of carbon fibres, some brand names and types will be given, for instance *IM7* or *T700*. The intention behind is to make it easier for the reader to find further information if needed.

2 State of research

The market for composite materials is growing worldwide and the demand for lightweight materials in different industrial applications is still on the rise. Consequently, recycling of composites is gaining importance. However, traditional recycling procedures are not directly transferable to composite materials. The state of research provides insights on recycling technologies of dry fibre fabric waste and end-of-life parts. It also introduces composites and their production, focussing on bulk and sheet moulding compounds. It begins with a thorough review of carbon fibres and their properties, followed by discussing testing procedures for single fibres.

2.1 Carbon fibres

A major question of material science is the clarification of the structure-property-relationships and process-induced variations of both, structure and properties. The production process is reflected in the mechanical and geometrical properties of carbon fibres and it is necessary to bear in mind the production route and the fibre history, for instance recycling steps, when trying to understand fibre and composite properties. Carbon fibres have some remarkable features. Depending on the manufacturing process, the fibres vary in shape and tensile behaviour. In particular, the cross-sectional areas can take a variety of shapes, from circular to kidney-shaped and even higher degrees of complexity. Independent of this, carbon fibres show the peculiarity of non-Hookean stress-strain behaviour, that is, at higher levels of strain, the Young's modulus increases [2]. Due to the significance for the mechanical properties of single fibres, the geometrical characteristics and the non-linearities are examined in detail in the work at hand. This is preceded by a brief excursion into the history of carbon fibres, their structure, manufacturing steps and fibre testing, laying the foundations for the understanding of the process-structure-properties-relation.

In the work at hand, the attention is drawn to PAN-based fibres as they have nowadays become the most important type of carbon fibres with a market share of approximately 96 % [3]. As only PAN-based fibres are investigated here, no further demarcation to carbon fibres made from rayon or pitch is made. It is worth introducing three abbreviations as PAN carbon fibres are available in different categories of performance: HM[1] abbreviates high modulus fibres; HT depicts high tenacity fibres and IM represents intermediate modulus fibres. In the work at hand, the widely used HT-PAN fibre PANEX 35 by *Zoltek Toray Group* is investigated on composite and single fibre level. Data taken from the according data sheet is listed in Table 2.1.
In terms of single fibre testing, some other PAN-based fibres are additionally examined to be able to discuss the finding obtained for PANEX 35[2].

Table 2.1: Properties of PANEX 35, also known as PX35 according to the technical datasheet [4]

Zoltek Toray Group PANEX 35 / PX35				
density *in g/cm³*	tensile strength *in GPa*	Young's modulus *in GPa*	diameter[2] *in µm*	carbon content *in %*
1.81	4.137	242	7.2	95

[1] In literature, the declarations *type-I* for high-modulus and *type-II* for high-tenacity fibres are also found but are not used here

[2] To simplify matters, "diameter" is indicated on the data sheet; presumably a circular fibre shape is assumed in this context.

2.1.1 Historic development and outlook

Carbon fibres had their first appearance at the end of the 19th century as filaments in incandescent lamps [5] but it was not until the 1960's that carbon fibres were commercially produced for structural applications [6]. A major breakthrough was the invention of high modulus fibre products by stretching the fibres during the heat treatment. To this point, cellulose precursors were primarlary being used and tensile moduli up to 500 GPa were contrivable by hotstretching of isotropic fibres at graphitisation temperatures [7]. The next key event occurred at the end of the 1960's when polyacrylonitrile (PAN) fibres were used as a precursor, resulting in a significant reduction of production costs. Research in the 1970's was dedicated to the clarification of the correlation between fibre production and fibre structure. Since the 1980's, and aided by the advance of transmission electron microscopy, structural investigations have been focussing on the atomic scale. Throughout the 1970's and 1980's, various companies improved their processing technology to enhance fibre properties, as summarised by Newcomb [8]. In terms of fibre properties, two, in those days opposing goals, were followed by tailoring the fibres to industrial needs: in one instance, designing high strength carbon fibres, in the other, maximising the Young's modulus. Since the turn of the millennium, research is also dedicated to the goal of creating fibres which combine high stiffness and high strength [9]. Furthermore, research on carbon fibres enters the field of sustainability by discussing the effects of recycling on fibre properties [10] and by returning to alternative, "green" precursors for fibre production [11]. Interestingly enough, the circle closes here in a sense as the above mentioned filaments for incandescent lamps were manufactured by the carbonisation of natural cotton fibres [5, 7].

To supplement the historical outline, it is worth taking a brief look at the development prospects of carbon fibres. Current challenges are hollow and tailored fibres which promise to be lighter than conventional carbon fibres and offer new perspectives in terms of a stronger interface by enabling additional frictional locking between fibre and matrix. Round [12–15] and complex shapes [14, 16] are already being researched and offer the potential for tailored composites. However, all types of hollow and complex-shaped fibres make fibre testing and composite design less predictable, as shown in the work at hand.

The most important challenge might be a further reduction in fibre production costs to open up new fields of application. Here, recycled fibres offer economic opportunities, provided that they keep their outstanding mechanical properties after recycling. Particular interest is shown in the automotive sector to meet future demands and regulation in terms of weight reduction, mileage and emission. The work at hand compares recycled and virgin fibres and introduces improved testing procedures to evaluate if recycled fibres meet the expectations of designers and engineers.
Nevertheless, the ultimate goal for research and development is to strive for the manufacturing of a new class of fibres, getting as close as possible to the theoretical values for tensile strength (106 GPa [17]) and the Young's modulus (1060 GPa [17, 18]) of graphite. Again, accurate testing procedures are needed to verify novel developments in fibre design. In addition, fibre testing can be used to guarantee the secure replacement of virgin fibres by recycled material.

2.1.2 Fundamental structure of carbon fibres

Mechanical properties and the manufacturing process are highly related to the structure of a material. Thus, it is advisable to give an overview of the process-structure-relations of carbon fibres [19]. This is particularly important when the material might undergo changes during recycling steps. Since this work is intended to contribute to a better understanding of the behaviour of virgin and recycled carbon fibres, this section starts with a review on today's understanding of the carbon fibre microstructure. Later, the fibre's geometry, its Young's modulus and its tensile strength are discussed in light of its structure.

Microstructure of carbon fibres

Many fundamental discoveries regarding the inner structure of carbon fibres date back to the 1980's, as advances in transmission electron microscopy, electron- and X-ray-diffraction enabled scientist to gain deeper insights into the atomic structure of carbon fibres. In particular, the comprehensive and scientifically recognised studies of Oberlin and Guigon [20–22] have contributed to a more precise understanding of the microstructure. Carbon fibres possess a turbostratic structure, consisting of smaller units, the so-called "basic structural units" (BSU) which are derived from the graphitic structure. Evidence of the turbostratic structure was provided by X-ray diffraction techniques as summarised by Johnson [23]. Graphite itself exhibits a hexagonal lattice structure as displayed in Figure 2.1a).

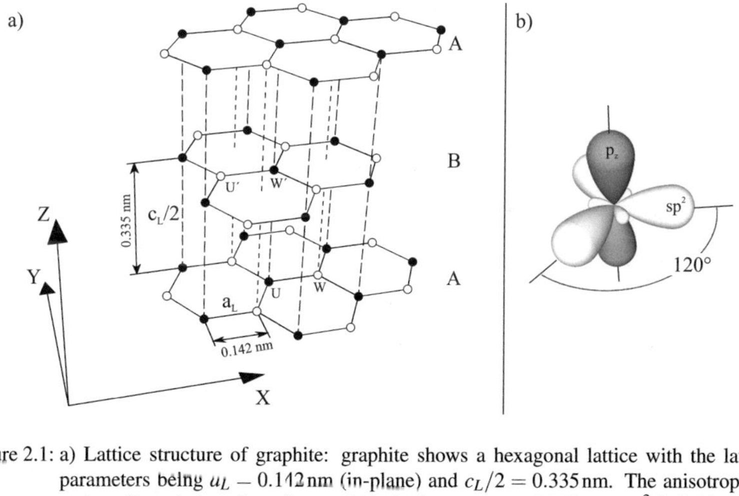

Figure 2.1: a) Lattice structure of graphite: graphite shows a hexagonal lattice with the lattice parameters being $a_L = 0.142$ nm (in-plane) and $c_L/2 = 0.335$ nm. The anisotropy of carbon fibres is based on the graphitic lattice structure; b) Planar sp^2 hybrid orbital found within each graphene sheet. *Partly based on* [17, 24]

The black and the white circles represents carbon atoms; the black ones possess a neighbour in the adjacent layer. The white ones only have a neighbour in every second layer. The interlayer spacing $c_L/2$ is approximately

0.335 nm, hence more than twice the distance that carbon atoms have within each graphitic plane. Given this significant difference in the lattice parameters in a_L- and c_L-direction, graphite and carbon fibres are anisotropic in their properties. For instance, electrical conductivity [25, 26], magnetic properties [27, 28] and mechanical properties [29] are clearly different in directions a_L and c_L. Layers A and B represent graphene sheets, with plane B being shifted by distance a_L in the X-direction. These sheets are connected by van der Waals forces, while covalent bonds are present within the honeycomb-like graphene sheets. The graphitic structure originates from the ability of carbon atoms to form sp^2-hybridised orbitals, see Figure 2.1b). These bonding conditions cause the high degree of anisotropy in both, regular graphite and carbon fibres. As Fitzer summarised, the ideal graphitic interlayer spacing of 0.335 nm is not achieved in carbon fibres but usually remains larger than 0.345 nm. Measurements for different types[3] of fibres were reported by Chae et al. [9] and Oya and Johnson [30]. Summarising these studies, the following values of the interlayer spacing were found, in ascending order: 0.337 nm for pitch-based fibre *K-1100*, 0.339 to 0.342 nm for HM fibres, 0.344 nm for an in-house product, 0.347 nm for fibres type *T700S*, *T800H* and *T1000*, 0.348 nm for *IM7* and 0.347 to 0.349 nm for *T300* [9, 30].

In general, the interlayer spacing is a function of heat control during the production, with higher temperatures of heat treatment resulting in decreased layer spacings [7]. Diefendorf and Tokarsky emphasised that to achieve high performance carbon fibres, a high degree of orientation is necessary to make use of the good properties in a_L-direction and to overcome disadvantages due to the poor properties in c_L-direction [31].

The graphite-like layers cluster in turbostratic structures which in turn arrange in the structure schematically shown in Figure 2.2. For detailed investigations of the structure and texture of carbon fibres, the reader may refer to [20, 21].

[3] In the following, trade names of some fibres are used as found in publications. This should facilitate an identification in case further information is seeked.

There is an consensus that an outer skin and an inner core exist in PAN based carbon fibres, as summarised by Newcomb [8]. The skin shows a higher degree of alignment and is more densely packed [32–34]. Modern microstructural investigations, by transmission electron microscopy (TEM), were supporting this hypothesis [35–37]. A model interpretation of the origin of a highly-ordered layer structure was presented by Bennett [38], who has also emphasised that the skin is more distinct in high-modulus fibres.

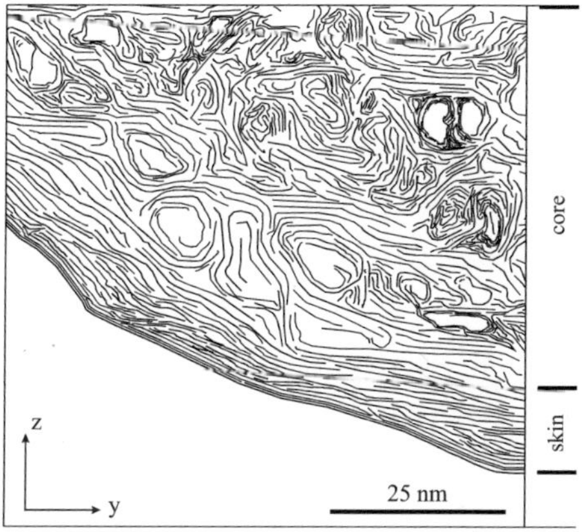

Figure 2.2: Schematic top view on a carbon fibre, revealing the structure of carbon fibres as described in literature [7, 20, 21, 38, 39].

Applying atomic force microscopy (AFM) on polished cross-sections taken from CFRPs, Wang and Hahn [40] found skin-core-structures in PAN-based fibres *T700* and *AS4* but not in the intermediate modulus fibre *IM7*. A skin-core effect was also described by Huang and Young [34] whose findings were supported by selected area diffraction patterns (abbreviated as SAD or SAED) and Raman spectroscopy. Applying a variety of methods, several studies suggested that the skin-core structure evolves during oxidation and stabilisation (Figure 2.3): nanoscale dynamic mechanical imaging [41], secondary ion mass spectrometry (SIMS) [42], energy-dispersive X-ray spectroscopy (EDX) [43],

Fourier-transform infrared spectroscopy (FT-IR) [43] and micro thermal analysis [44] all ascertained a correlation between the conditions during oxidation and stabilisation and the resulting manifestation of a skin-core structure. The radial structure of PAN precursors and consequent difference between skin and core region was described in [45]. Morris et al. [46] concluded that fibres with diameters as small as 2.5 µm showed less skin-core heterogeneity and a potential for better mechanical properties.
In addition to skin and core, the term "sheath" [23, 38, 47, 48] was introduced in literature. Bennett [38] quantifies the different regions: for HM fibres, the skin was found to be a region of large, highly orientated crystallites with a size of 10 nm and a layer thickness of approximately 100 nm. It was further defined that the sheath is a fully-stabilised transition zone containing voids, adjacent to the under-stabilised core of densely packed crystallites. Crystallite sizes and the spacing of graphitic planes in light of heat-treatment as well as the resulting skin and core were also discussed in [23]. In contrast, the skin was estimated to account for 20 % of the total cross-sectional area of fibres in [21].
In the work at hand, no further distinction between skin and sheath are needed. The topic is simplified by assuming that an ordered skin and less ordered core region can be expected.

In general, Loidl, Paris and their co-workers [49–51] agreed that, commonly, the skin region shows higher orientation than the core. However, the investigations partly contradicted, stating that the effect could be reversed depending on fibre type and heat treatment.
Despite all agreement in the scientific community, there are recent publications, particularly [52–54], which gave reason to believe that the structure of carbon fibres has not been fully understood yet. These authors found Young's moduli of nanocrystallites exceeding the tensile modulus of graphite by up to 10 %. This contradicts the general understanding, that the maximum Young's modulus of carbon fibres is supposed to be the modulus of graphite and could not be exceeded if carbon fibres consisted of nothing but basic structural units.

Nevertheless, the work at hand relies on the currently accepted understanding of the structure of carbon fibre. Equipped with these insights on the general structure of carbon fibres, the manufacturing process is explained and the process parameters influencing the structure and properties of the fibres are unrolled.

2.1.3 Manufacturing, process parameters and heat treatment

Carbon fibre properties are strongly related to the processing steps and the accompanying temperature control [8, 55, 56]. Moreover, the temperature control determines whether HM CF (high modulus carbon fibres) or HT CF (high tenacity carbon fibres) are produced. HM fibres are not only superior in terms of the Young's modulus but also up to ten times more expensive than commodity HT fibres [19]. In this section, a brief overview of the productions steps for PAN based fibres is given and literature concerned with the resulting microstructure and properties is discussed. These fundamentals are necessary to better understand the structure-property relationships and the results of the mechanical characterisation of single fibres. Comprehensive reviews on the different processing steps of carbon fibres can, for instance, be found in [7, 19, 55, 57–59]. The findings are summarised and discussed in the following section. An explanatory illustration (Figure 2.3) of the process steps can be found at the end of this section. Detailed chemical insight is found in [60] and definitions are provided by IUPAC [61]. The aim of the processing steps is to increase the carbon content of the fibres and to align the synthesised graphite-like layers (BSU) along the fibre axis [31, 57]. - 1600 °C

Production of PAN precursor and precursor fibres

The PAN route starts with acrylonitrile, which is polymerised to polyacrylonitrile at temperatures of 40 - 60 °C [58]. Given the high initial carbon content of 68 wt.-%, PAN qualifies as precursor for the production of carbon fibres but several steps are needed to remove the other elements. The first step is spinning at 0 - 90 °C to obtain PAN fibres [58]. As summarised by [62, 63], different spinning procedures are available. The spinning process has a decisive influence on the later fibre properties: the quality of the precursor is reflected in the final carbon fibre [57]. For instance, Moreton and Watt observed [64] that clean room conditions during spinning of PAN fibres lead to carbon fibres, which had a significantly higher tensile strength than fibres produced in customary condition. The same conclusion was drawn by Johnson and Thorne in an earlier work [65]. Instead of clean-room conditions, the flaw population of the precursor was correlated with tensile behaviour of the corresponding carbon fibre. Serkov et al. [66] noted that the fibre quality and thus the mechanical performance is predefined in the production of the precursor. Particularly, increased porosity and reduced uniformity of fibre diameter were expected to have a negative influence on fibre properties.

The correlation between precursor condition and the resulting carbon fibre quality was also discussed in [67–69]. A profound review of the same correlation, focussing on post-spinning of PAN fibres, was given by Mittal et al. [70]. Carbon fibres made from PAN precursors with high molecular weight offer the potential of a decreased pore content and increased mechanical performance [46]. High precursor density and appropriate fibre tension were also listed as beneficial in to achieve high quality fibres [71].

Stabilisation and carbonisation

The PAN polymer fibres are already anisotropic and have a preferential direction. Stabilisation and carbonisation (Figure 2.3) aim at increasing the carbon content of the fibres. To do so, the polymer fibres are first oxidised to the so-called oxidised or stabilised fibres. The stabilisation takes place at elevated temperatures of 200 - 300 °C and under applied mechanical tension [57, 58]. Carbonisation is run at temperatures of roughly 1000 - 1700 °C and in inert atmosphere, commonly nitrogen. The chemical reactions during stabilisation

and carbonisation cross-link the polymer and result in a graphit-like structure and an increase in density [24, 56, 57]. The density increases due to the elimination of volatiles [57], as displayed by Fitzer et al. [56]. According to [71], the stabilisation rate and the extent of applied tension are important factors to yield high quality fibres.
Heat treatment is decisive in terms of the structure of the carbon fibres. Carbon fibres with maximum tensile strengths of up to 7 GPa are obtained when carbonisation is run between 1400 - 1600 °C, as at higher processing temperatures, crystallite size and void content increase [24, 72–74]. Stabilisation and carbonisation lead to an increase of the carbon content to at least 92 wt.-% [61]. Further information concerned with models and critical discussions of the chemical reaction during thermal stabilisation, particularly cyclisation, are found in [75] and [76].

Graphitisation

If the aim is to produce high-modulus carbon fibres instead of fibres with maximum strength, an additional step is needed. To achieve high modulus fibres, a last heating step is applied, raising temperatures to approximately 2200 - 3000 °C [57, 58]. As a consequence, HM fibres exhibit a lower tensile strength than HT fibres. Graphitisation is expensive due to the very high processing temperatures and the requirement of using argon as inert gas at temperatures above 2000 °C. Structural changes occurring during graphitisation lead to higher crystallite sizes and degrees of orientation, resulting in a higher Young's modulus [24]. Ozbek et al. [77, 78] investigated the influence of hot stretching carbon fibres and found the following trends: higher temperatures and higher stretching stresses lead to a significant increase in apparent crystallite size and alignment, expressed by an angle of the preferred orientation parameter closer to 0°. Both, the increase in temperature and stress during hot stretching lead to a significant increase in the Young's modulus of the fibres, accompanied by a reduction in fibre diameter. As a consequence, higher orien-

tation is also correlated with a smaller fibre diameter. An increase in density with these parameters was also observed [79].
Due to the resulting costs for the HM fibres, the graphitisation step is only applied if a high modulus end product is to be produced [24].

Surface treatment and sizing of carbon fibres

The final step includes surface treatment and sizing processes. The carbonised or graphitised fibres are oxidised to improve the adhesion of the fibre to the sizing or the matrix and to ensure load transfer between matrix and fibre [80, 81]. Where appropriate, chemical sizing is applied to protect the fibres from being damaged due to wear during handling and processing, to adjust their wetting properties and to fixate single fibres into rovings [80, 82, 83]. Summarising, surface treatment and sizing are intented to tailor the fibre/matrix-interface, to fixate and protect the roving and serve as coupling agent for an optimised fibre-matrix adhesion [80].
The carbon fibres used in this work, PANEX 35, are covered by approximately $1.5 \pm 0.5\,\%$ epoxy sizing but no detailed contents are declared. After surface treatment and packaging, carbon fibres are ready to be used as reinforcement in polymers (CFRP).

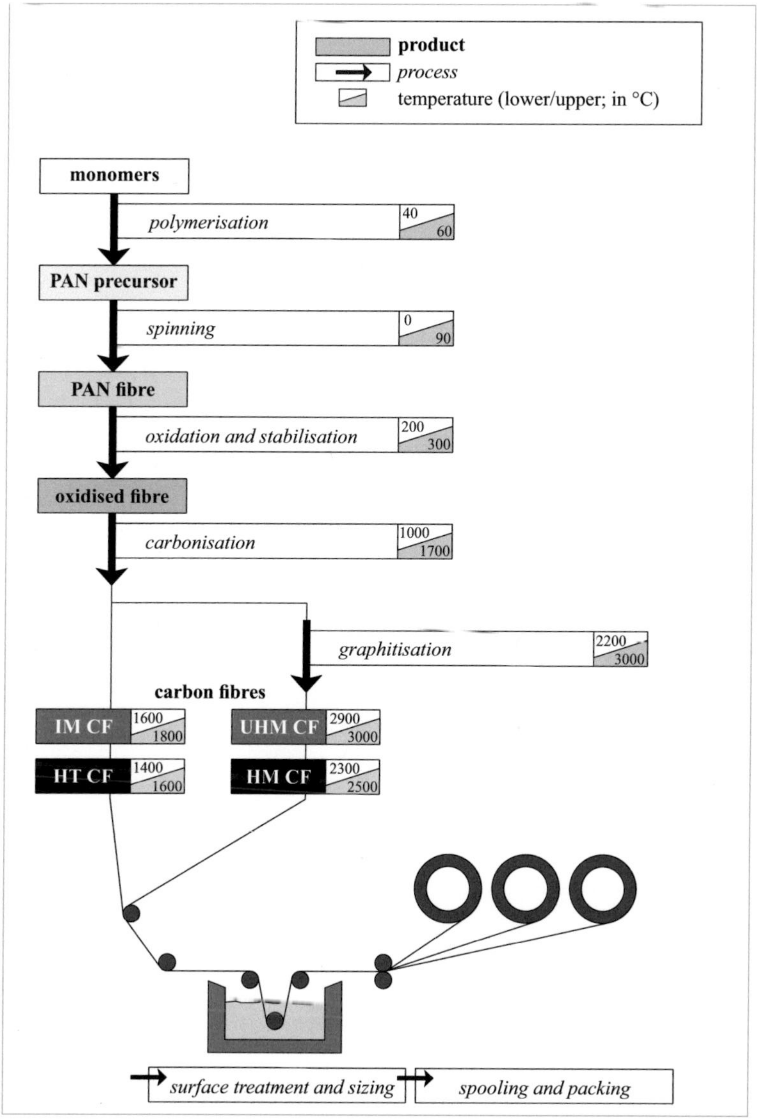

Figure 2.3: Scheme of the production of PAN based carbon fibres from the monomers to HT and HM fibres. The process chart and the approximate process temperatures result from sources [7, 57, 58].

2.1.4 Fibre testing

Enhancements in the field of single fibre testing are one one of the major experimental aspects of the work at hand. In order to motivate the scientific question how to promote single fibre testing procedures, this section is dedicated to reviewing literature in the field of single fibre tensile tests and how to derive fibre properties.

Motivation of single fibre tensile testing

The testing of single fibres is gaining importance due to several reasons. Knowing the properties of single fibres leads to a better understanding of composites and enables improvements on simulation models. Here, accurate values for the cross-sectional area, the tensile properties and the Poisson's ratio are of interest to refine micromechanical models. The composite market and thus the necessity to recycle composites is still growing. Single fibre properties are particularly important when the recycling impact on a composite is to be investigated as in [10, 84]. Without profound knowledge about possible impacts of recycling on fibre properties, it is difficult to convince users to use recycled materials in their future design concepts. Single-fibre testing, if executed correctly, presents an opportunity to understand the impact of recycling on carbon fibres, typically accompanied by full-scale and coupon testing.
If the aim is to improve the determination of Young's modulus E of technical fibres, a glance at equation 2.1 is enough to get an idea which quantities should be determined with increased accuracy.

$$E = \frac{\Delta\sigma_T}{\Delta\varepsilon} = \frac{\Delta F / A_0}{\Delta\varepsilon} \tag{2.1}$$

Here, $\Delta\sigma_T$ is the change in engineering stress, $\Delta\varepsilon$ the change in strain, ΔF the change in load and A_0 the initial cross-sectional area of the fibre.

It is apparent that the determination of the Young's modulus benefits from improvements of either the measurement of the force, the strain or the cross-sectional area. The following literature review is motivated by Equation 2.1. Theory concerning load measurements is not discussed here. Load measurements were not under investigation in the work at hand. The second parameter, the cross-sectional area, is debated in Section 2.1.5. The third parameter, determination of strains on fibre level, was addressed in several works and is discussed in the following.

The early years of testing cotton

To understand the motivation to develop methods in the field of single fibre testing, it is worth taking a brief look at the historical development of the test methods and their background.
In the field of cotton, different testing devices have been developed over decades. Already in 1937, different working principles for single fibre and yarn testing were reviewed in [85] and date back to 1863 [86]. The importance of testing fibres, no matter if cotton or technical fibres, is thus long realised. The basic principles found in [85] were also used to design two popular fibre bundle testers, the Pressley tester [87] and the stelometer. Both are shown and described in ASTM D1445/D1445M-12 [88] and allowed for systematic and repeatable investigations on cotton properties. The Pressley tester was used in several historic studies to determine the strength of cotton, for instance in [89, 90]. Systematic and reproducible testing of cotton was of special interest to account for fluctuations of the quality of cotton harvested in different seasons and regions [91]. A current overview of cotton properties and state-of-the-art testing devices was given by [92].

From cotton bundles to single fibres

Fibre bundles are much easier to see and to handle than single fibres, which makes testing efficient. However, fibre bundle tests show major disadvantages, which were first discovered when testing cotton and then later glass fibres: it is hardly possible to determine the actual cross-sectional area of the bundle, as voids exist between single filaments. Likewise, it is highly debatable if the external load is distributed uniformly onto the fibres. In particular, after failure of the weakest fibre within the bundle, the adjacent fibres have to carry a higher load all of a sudden [93, 94].

Cowking et al. [95] combined fibre bundle testing on glass fibres with acoustic emission. Relating each load drop with an audible signal, conclusions about the strength of individual fibres could be drawn. Nonetheless, it was mentioned that the shock caused by the fracture of one fibre could initiate further breakages [95]. The resulting non-uniform load distribution leads to a non-linear stress-strain-curve and progressive failure [96]. The latter is considered less critical for the determination of the Young's modulus and mainly relevant in terms of tensile strength, but the former will increase measuring errors in the determination of the Young's modulus. The non-linear behaviour and the progressive failure mechanism are both attributed to the statistical strength distribution of fibres [96].

This systematic source of error in the experimental method reflects in several studies on cotton. For instance, Orr et al. [97] compared seven different types of cotton. When testing cotton bundles, tenacity was on average 58 % lower compared to single fibre tests. In terms of elongation, both experiments varied up to 27 %. The importance of considering true gauge lengths to account for slip effects of the fibres was also emphasised. Gonsalves [98] observed that cotton strength in bundle tests was reduced by 33 % compared to single fibre testing conducted using a device called Heim Tester. Furthermore, it was made use of the later described vibroscopy method to determine cross-sectional areas of the fibres. Given irregularities in fibre bundles, procedures using single fibres are recommended. Sasser et al. [99], also dealing with cotton, emphasised this finding and found that the average single fibre tensile strength was the upper limit of the strength determined for bundles. Delhom et al. [100] found linear relations between bundle and single fibre tests on cotton with gauge lengths of 3.175 mm and 12.7 mm in terms of strength and elongation at failure. Single fibre properties were significantly higher, particularly for a shorter gauge length.

Hill and Okoroafor [101] tested lubricated and as-received bundles of glass and aramid fibres and pointed out another significant downside of bundle testing: friction between individual fibres in a bundle is associated with a decrease in tensile strength and breaking strain of 10 to 20 %. However, a variation of the Young's modulus between the two states was not found. Research findings regarding the effects of friction in non-impregnated carbon fibre bundle testing was not available.

Hughes et al. [102] discussed the correlation between single fibre and bundle strength: the effective bundle strength is determined by the weaker fibres within the bundle and as a consequence, the bundle strength is less than the average of single fibre strengths within the bundle. Coleman [103] deduced that the fibre bundle strength would only equals the single fibre strength if all individual fibres had the same strength. As real fibres show a statistical distribution of individual fibre strengths, the fibre bundle strength is less than that the average of individual fibres. Comprehensive examinations of statistical theories of fibre bundle behaviours are found elsewhere [93, 94]. For the reasons listed here, bundle tests were not carried out in the work at hand.

While an unequal contribution of single fibres in a bundle are a disadvantage of bundle tests, saving time is advantageous [97]. To counteract the disadvantages of bundle tests, single fibre testing devices have been developed over the years. Higher throughput and easier handling were achieved by the Mantis tester [104] and later the semi-automatic FAVIMAT+, for instance applied in [105]. FAVIMAT+ is nowadays also applied for technical fibres, for instance for carbon [106, 107] and glass fibres [108].

Not for cotton but hair, the company *Dia-Stron Ltd* entered the market of fibre testing[4]. As these testing procedures evolved in the field of (cotton) textiles and hair, another factor has to be considered: technical fibres such as glass and carbon vary by an order of magnitude in the Young's modulus compared to cotton. Thus, transferability of traditional textile testing procedures to technical fibres cannot be taken for granted. As the literature presented here also reveals, the focus of cotton testing is to determine tensile strength while for technical fibres, stiffness is equally important. It is therefore essential to avoid possible sources of error when investigating fibre stiffness [109, 110]. Therefore, the following paragraph is devoted to review the corresponding standards and approaches for enhanced single fibre testing techniques.

Test standards and testing procedures for fibres

Starting with standards for cotton testing [88], standards for technical fibres have evolved. They deal with both, determination of cross-sectional areas [1] and tensile properties. Procedures for tensile testing are found in standard ISO 11566:1996 [111], reapproved in 2016, in ASTM standards D3379-75 [112], withdrawn in 1998, and in C1557-14 [113], reapproved in 2014.
Both active standards focus on a method where the fibres are glued to mounting tabs prior to testing. Commonly, cardboard frames are used as mounting tab template. Thus, the method is henceforth referred to as cardboard frame method. Details of the procedure are given in [113] and are briefly discussed here: cardboard frames are prepared and either a hole or slot is pre-cut in the central region of the frame. The fibre is aligned and glued to both ends of the mounting tap. The slot/hole size together with the distance between the glue spots defines the gauge length. The cardboard frame can then be clamped into a suitable universal testing device. The cardboard is cautiously cut and the fibre is then loaded until failure.

[4] https://www.diastron.com/about-us/company-history/ - 07.05.2019

However, the cardboard frame method is dissatisfying in several aspects: it is time-consuming and reproducibility is hard to maintain: pre-tensions due to the gluing are impossible to measure, the exact gauge length varies as a function of gluing precision and deformations of the whole systems have to be expected due the low stiffness of the cardboard frame and the glue spots. To counteract these effects, several sets of fibre lengths have to be tested to account for the system compliance. The load train system and the specimen-gripping-system contribute to the system compliance. A procedure to determine the system compliance is found in both standards [111, 113]. The importance of compliance correction was for example addressed in an experimental work with ceramic fibres [114]. The reliability of compliance corrections was discussed in several works: Li and Langley [115] recommended to take the fibre cross-sectional area into account. Linear regression of their data set was significantly improved when the volume is taken into account instead of following standard compliance corrections focussing on fibre length [111]. Non-Hookean material behaviour [116] and fibre slippage [117] are also reasons of faulty compliance corrections.

Summarising, the cardboard frame method has the advantage that it can be used for all types of fibres and in any universal testing machine without the need to install elaborate testing devices. However, its downsides are manifold: compliance corrections have to be carried out, preparation is time-consuming and prone to inaccuracies. In-situ determination of the cross-sectional area is complicated by the presence of the cardboard frame.

To overcome the time-consuming fibre preparation needed for the cardboard frame method, several aforementioned single fibre testing devices have been developed, for instance the FAVIMAT+ by *Textechno* and *LEX810* by *Dia-Stron*. In their commercially available versions, direct strain measurement is not implemented and thus these devices still lack the capability to cope with issues caused by the system compliance. To determine the Young's modulus, fibre clamping lengths have to be varied whenever the type of fibre or gripper material is changed.

For these types of single fibre testing devices, the fibres can be mounted in different ways: they can be clamped with metal grippers coated with a polymer layer to protect the fibre surface and prevent crack initiation within the clamping regions as this would lead to invalid tests. These polymer layers are less stiff than the fibres and will deform when a load is applied as demonstrated in Figure 2.4. Alternatively, fibre ends can be embedded in resin, which might also deform when loaded.
Analogously, deformations of the system can occur when applying the cardboard method. Figure 2.4 demonstrates both, in a), the fibre is clamped by polymer-coated metal grippers, in b), the fibre is glued to cardboard frames.

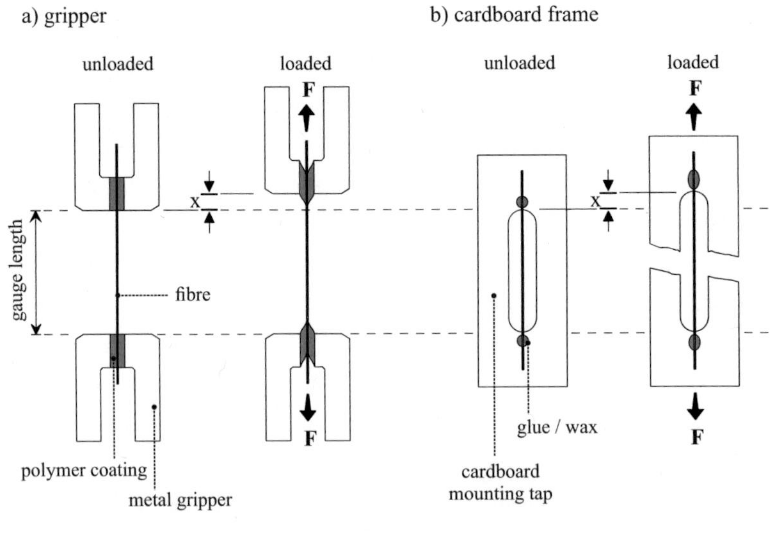

Figure 2.4: Schematic representation of the deformation of the different gripping systems when the tested fibre is loaded

In Figure 2.4 a) and b), the left image shows the initial state, the right image the loaded state when a force F is applied during the tensile test. The coating and the glue can be deformed when tension is applied. In addition to the elongation of the fibre, the system itself is then strained by a distance x.

Conventional testing devices measure fibre elongation by logging the position of the machine's crosshead. Due to the finite rigidity of all testing setups, strains as given by crosshead displacement are misinterpreted and should be replaced by direct measurements on the fibre. The additional movement by distance x will cause erroneous readings. By applying direct measuring techniques, these errors can be prevented.

As elucidated later, direct strain measurements make compliance corrections obsolete, overcoming the associated disadvantages. While indirect strain measurement means that the displacement of the testing machine's crosshead is being observed, direct strain measurements acquire strain information on fibre level.
This work aims on closing this gap and launching an enhanced method to handle single fibres and measure their mechanical properties as already presented by the author in [109, 110]. To do so, this literature review is extended to testing devices and procedures applied for technical single fibres.
Direct elongation measurements by attaching flags onto the fibre are mentioned in ASTM standard C1557-14, but no further specification is given. Neither the flag material and proportions, nor the equipment to track the positions of the flags are specified. For reasons not stated, applicability to fibres with diameters below 5 µm is doubted in the standard. In addition, it is recommended to determine the diameter in-situ [111] or subsequent to failure [113], the latter considerably increasing the total duration of the testing procedure.

Digital image correlation

A comprehensive overview on testing materials in tension was given in [118]. Among the techniques applied on macro-scale specimens are mechanical extensometers, strain gauges and digital image correlation. For obvious reasons, neither extensometers nor strain gauges can be used in single fibre testing as the fragile fibres could not withstand the load induced by these devices. Instead, non-contacting methods such as digital image correlation (DIC) or laser-based methods are the most suitable alternative.

As a consequence, digital image correlation is the method of choice for correctly measuring strains on fibre level. Thus, a few remarks about DIC are pointed out here. In material testing, DIC is a powerful tool to determine local and global strains at the same time.

It can be applied to different testing modes, for instance in bending, compressive and tensile mode. To apply digital image correlation, unique optical features are needed on a specimen. Commonly, these features are applied by spray-coating the specimen with a stochastic speckle pattern. In a software, rectangular sections, so called facets, are applied. These facets are tracked over time and displacements and strains can be evaluated. Applications and literature in this field are manifold and comprehensive overviews were provided in [119–123]. The implementation in the GOM[5] software GOM *Correlate Professional*, was described in [123]. It was to some extent based on the groundwork by the founder of the company GOM [124]. Instead of camera-based systems, laser techniques can be applied as presented by Reder et al. [125]. The non-contacting technique presented in [126] for bulk materials was transferred to fibres. The importance of direct strain measurements was emphasised. It was furthermore argued why non-contacting methods are reasonable. Concluding, it was stated that the method not only overcomes the issue of compliance corrections but also offers the potential of fibre testing at elevated temperatures [125].

So far, all image correlation techniques on single fibre level rely on flags or other markers attached to the fibre as recommended in [113]. This is due to limitations in the optical resolution of camera and video systems. Fibres themselves are objects too thin to yield a recognisable optical pattern.

The feasibility of applying DIC on fibres and wires was thoroughly studied by Hendrickx et al. [117], Depuydt et al. [127] and the author of the work

[5] Gesellschaft für optische Messtechnik mbH, Braunschweig, Germany

at hand [109, 110]. Both works of Hendrickx and Depuydt et al. can be summarised as follows. Steel wires ($d = 30\,\mu m$) as well as bamboo and flax fibres were tested using the cardboard frame method with a clamping length of 50 mm. Results obtained by indirect strain measurements given by the crosshead displacement and considering the system compliance and by direct strain measurements using digital image correlation were compared. For DIC, one white flag of 3 mm in diameter was attached to each end of the fibre and sprayed with a black stochastic pattern. A DIC system with a spatial resolution of 96 dpi was used. In general, satisfactory agreement between both methods was found. The advantage, that direct strain measurements do not demand compliance corrections, was highlighted as this allows for significant time savings. This was particularly beneficial as they also found that correction methods for the system compliance underestimate the fibre stiffness [117]. It was assumed that the system compliance might not be constant and might be affected by slippage. Nonetheless, disadvantages of this approach to direct strain measurements on single fibres were listed in [117, 127]: The digital image correlation can be affected by out-of-plane movements of the fibre or the tracking flags as well as by insufficient light conditions. Camera resolution was a significant factor and was relatively low in [117, 127].

Direct strain measurements were also evaluated using video extensometry. Coimbra et al. [128] used video extensometry to determine strains and calculate the Young's modulus of ceramic fibres by attaching paper flags to the fibres. However, using paper flags as markers demands for a high degree of alignment. Ambient conditions and draught may cause fibre movement and disrupt the strain measurement. Adusumalli et al. [129] tested direct strain measurements with cellulose fibres by applying ink on the fibre and tracking an intensity pattern. For strains higher than 1 %, a good correlation with data obtained indirectly by crosshead displacement was found. Still, using the cardboard methods remained a time-consuming technique and their video camera with a resolution of 1.3 megapixel might have been too inaccurate for good measurement quality. Burgert et al. [130] tested video extensometry on wood

fibres. In their approach, markers on the gripping system in the proximity of the fibre were tracked. However, this cannot overcome compliance issues caused by gluing the fibre to the clamps.
The approaches of direct strain measurement presented so far were all applied to natural fibres or metal wires. A direct transferability to carbon fibres is not guaranteed given significant differences in cross-sectional areas and stiffness. Furthermore, the resolutions of the camera systems stated in [117, 127, 129, 130] were rather low.
With regard to testing the mechanical properties of single carbon fibres, equipment and services are available on the market. Examples are the aforementioned FAVIMAT+ by *Textechno* and *Dia-Stron's* LEX820/LDS0200. Being semi-automatic, they overcome several disadvantages of the manual cardboard method and are time-efficient. Nonetheless, they do neither provide a solution to determine non-circular cross-sectional areas nor strains directly on the fibre. This results in scientific considerations, which alternative procedures to determine both, cross-sectional area and strain, can be designed.

2.1.5 Examination of cross-sectional areas of carbon fibres

The manufacturing of carbon fibres does not only have an impact on the fibre structure and the mechanical properties but also on the shape and size of the cross-section. The latter plays a key role when the fibre geometry is to be determined as accurately as possible prior to mechanical tests and when the Poisson's ratio is investigated (Section 3.1.4). A brief summary of the state of research concerned with fibre shapes and the determination of the same is given in the following.

Geometric derivation to obtain carbon fibre shapes

To understand the procedure developed in this work, a mathematical definitions of an ellipse and a convex set are needed. The microsections of carbon fibres, when viewed perpendicular to the fibre axis, can resemble oval shapes, as shown in Figure 2.5a). The contour can then be traced as depicted in Figure 2.5b). In turn, a convex oval can be approximated by a regular elliptical shape, shown in Figure 2.5c).

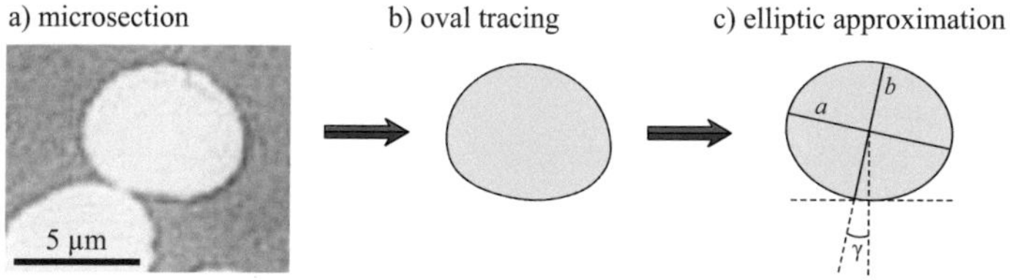

Figure 2.5: Approximation of the fibre shape: oval shapes of carbon fibres can be approximated by ellipses.

The area of an ellipse is defined by equation 2.2 [131]:

$$A_{ellipse} = a \times b \times \pi \tag{2.2}$$

The semi-major and semi-minor axes are labelled a and b, respectively, and have to be determined to approximate the cross-sectional area of an oval shaped carbon fibre.
To apply laser diffraction techniques, a boundary condition is the convexity of the fibre shape. Therefore, a brief definition of the term "convex set" is useful. Convex sets are, in consideration of Figure 2.6, defined as follows. A set S is defined as being convex if and only if for all arbitrary points P' and P'' in S and all points X' on the line segment connecting P' and P'', the points X' also belong to the set S [132, 133]. If any point X' does not belong to the set, the set is non-convex. Figure 2.6 shows a convex set $S1$ on the left and a non-convex set $S2$ on the right.

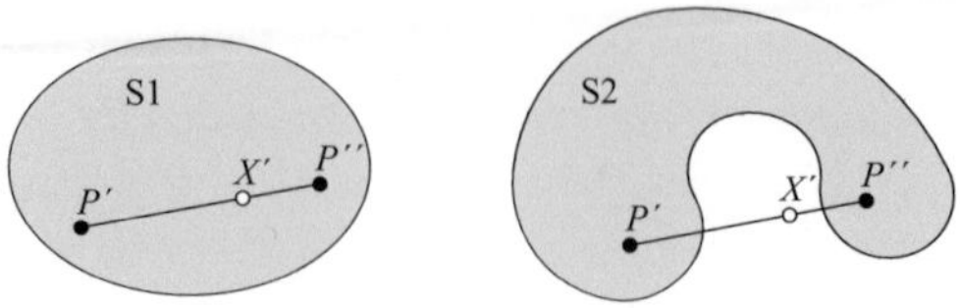

Figure 2.6: A convex set (left) and a non-convex set (right) to deduce the convexity of ovally-shaped carbon fibres.

Strictly speaking, this condition must apply if laser diffraction or laser scanning techniques are used to determine the fibre geometry. Apparently, non-convex shapes lead to undercuts that laser techniques cannot determine but instead, overestimate the cross-sectional area. It must be assumed that slight deviations from a strictly convex shape lead to a "smoothening effect" (see also Figure 3.32). These deviations are, for instance, caused by the roughness of the fibre. Evidently, the same difficulties arise when using state-of-the-art techniques such as vibroscopic methods or lateral inspections with scanning electron or light microscopes.

Typical fibre shapes and their classification

A review of literature demonstrates the vast extent of shapes the cross-sectional areas of carbon fibres can assume. Carbon fibres can be assigned to four major categories, shown in Figure 2.7: circular (a), oval and elliptical (b), hollow or other complex shapes (c) and kidney-shaped (d). For comparison, two real examples of fibres of type PANEX 35 are also included in Figure 2.7, namely (e) and (f). Apparently, real fibre shapes can be approximated by the four categories shown in Figure 2.7 (a) to (d). The upper row contains convex shapes and qualifies for measurements by means of laser diffraction. For the non-convex shapes displayed in the bottom row, inaccuracies in the determination of the cross-sectional area must be expected. Undercuts and particularly cavities of hollow fibres cannot be measured by laser diffraction; the area would be overestimated.

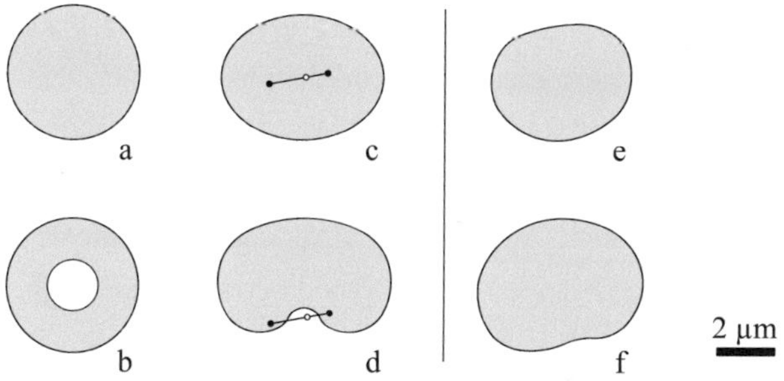

Figure 2.7: Classification of fibre shapes: according to literature, the shapes of the cross-sectional areas of carbon fibres can be assigned to the four main categories *a* to *d* (2.2). Images *e* and *f* represent actual fibre shapes derived from microsections of PANEX 35 fibres, manufactured by *Zoltek Toray Group*. The upper row shows convex fibre shapes, the bottom row non-convex fibre shapes as symbolised by the line and explained in Figure 2.6.

An overview of publications presenting either SEM or microscopic investigations on fibres is given in Table 2.2. The clustering in Table 2.2 is based on the categories given in Figure 2.7. In addition to the bibliographic data, the respective precursor used is declared. The major investigation of the work is assigned to one or two of the following topics: tensile properties, compressive properties, composite properties, microstructure, (fibre manufacturing) process, application, method development or chemical properties.

The references given in Table 2.2 are to be understood as examples and do not claim to be exhaustive. For instance, further possible complexly shaped carbon fibres are presented in [134]. The decisive factor is rather the realisation that due to the large number of different fibre shapes, great care has to be taken when seeking an appropriate tool to investigate the cross-sectional areas.

For the fibre shapes presented in category (a), a calculation of a circular diameter based on an apparent diameter seems appropriate. Furthermore, laser diffraction and side views in SEM qualify equally. In the work at hand, a procedure to approximate the cross-sectional areas of fibres from category (a) and (c) is introduced. Fibres from category (b) and (d) have to be handled with great care. None of the in-situ methods presented in BS ISO-11567 can be applied correctly; only optical and electron microscopy investigations of polished faces can be indicative. However, they have to be run subsequent to tensile tests.

It is worth noting that the cross-sectional shapes of carbon fibres are predetermined by their precursors. Different precursor shapes are presented in literature, for instance kidney-shaped [19, 135] and circular shapes [19].

Table 2.2: Literature review of the cross-sectional areas of carbon fibres

author	year	citation	precursor	major topics
category a: circular				
Naito et al.	2008	[136]	PAN, pitch	tensile properties
Naito et al.	2012	[137]	PAN, pitch	tensile properties
Naito et al.	2017	[138]	PAN, pitch	compressive properties
Kumar et al.	1993	[139]	PAN, pitch	compressive prop., microstructure
Endo	1988	[140]	pitch	microstructure
Hughes	1986	[141]	n.d.*	tensile properties
Cho et al.	2003	[142]	pitch	microstructure, process
Tzeng & Chang	2001	[143]	PAN	application
Vezie & Adams	1990	[144]	PAN, pitch	microstructure
Liu et al.	2012	[145]	n.d.	composite properties
Kobets & Deev	1998	[146]	n.d.	microstructure, tensile properties
category b: complex, for instance hollow or cross-shaped				
Curtis & Travis	1999	[13]	PAN	process
Liu et al.	2015	[12]	PAN	process
Wang et al.	1998	[15]	pitch	process, microstructure
Xie et al.	2011	[147]	PAN	process, microstructure
Park et al.	2004	[148]	pitch	composite properties
Gulgunje et al.	2015	[14]	PAN	process, microstructure
Hunt et al.	2012	[16]	polyethylen	process
Liu et al.	2012	[145]	n.d.	composite properties
McMahon	1973	[134]	n.d.	tensile properties
Eom & Ryu	2010	[149]	pitch	chemical properties

** n.d. = not declared*

Table 2.2 (continued): Literature review of the cross-sectional areas of carbon fibres

author	year	citation	precursor	major topics
category c: **elliptical or oval**				
Naito et al.	2008	[136]	PAN, pitch	tensile properties
Huether et al.	2019	[110]	PAN	method development
Kobets & Deev	1998	[146]	n.d.*	microstructure, tensile properties
Saburow et al.	2017	[150]	PAN	composite properties
Li et al.	2017	[151]	n.d.	composite properties
Paris et al.	2001	[152]	PAN, pitch	microstructure
Fitzer	1989	[7]	PAN, pitch	microstructure, process
category d: **kidney-shaped**				
Naito et al.	2008	[136]	PAN, pitch	tensile properties
Naito et al.	2017	[138]	PAN, pitch	compressive properties
Endo	1988	[140]	pitch	microstructure
Hughes	1986	[141]	n.d.	tensile properties
Daumit & Ko	1986	[153]	PAN	process
Kumar et al.	1993	[139]	PAN, pitch	compressive prop., microstructure
Marcuzzo et al.	2013	[154]	PAN	process, chemical properties
Tzeng & Chang	2001	[143]	PAN	application
Kobets & Deev	1998	[146]	n.d.	microstructure, tensile properties
Johnson & Thorne	1969	[65]	PAN	tensile properties

* *n.d. = not declared*

2.1.6 Procedures to measure the cross-sectional area of carbon fibres

The previous section qualitatively discussed the state of research concerned with the shapes of carbon fibres. Ways how to quantify the cross-sectional area of carbon fibres are covered below.

Determination of cross-sectional areas according to the standard

Standard BS ISO-11567 [1], dedicated to the cross-sectional areas of carbon fibres, shall be summarised here. It offers five methods to determine the cross-sectional area of carbon fibres. It is emphasised that the term diameter can refer to a *true* diameter in case of a perfectly circular fibre and to an *apparent* diameter, if only the side view of a non-circular fibre is observed. The fact that this is explicitly stated in the standard indicates that circularity should always be discussed and verified.
The methods recommend in BS ISO-11567 [1] are summarised in the following, accompanied by statements to what extent they are applied in the work at hand.

A An average diameter can be calculated when the linear density is known. As this method is used for bundles or yarns and gives an average fibre diameter, it is not suitable for the investigation followed in the work at hand.

B Optical microscopy (OM) can be used to determine the apparent diameter of carbon fibres but is not recommend for fibres thinner than 10 μm due to diffraction effects.

C Fibre bundles can be embedded in a resin block. A polished face, perpendicular to the fibre direction, can then be investigated by means of optical or electron microscopy. OM is recommended for fibres above a threshold of 10 μm; SEM for smaller diameters. In addition to the resolution of the microscope, the correct vertical embedding is of utmost importance for the quality of the measurement. Here, both measurements were applied to allow for a comprehensive comparison.

D Laser diffraction can be used to determine an apparent fibre diameter. It is recommended to measure the diameter at multiple positions along the fibre and at different angles with respect to the fibre axis. Laser diffraction is the main technique used in the work at hand and both aforementioned limitations are taken into account, as explained in Sections 3.1.2 and 3.2.

E Analogous to optical microscopy, scanning electron micrographs can be taken as side views. SEM was used in the work at hand to obtain comparative values.

$\hookrightarrow$ To date, only method C is capable of considering the real cross-sectional area of a fibre. However, it cannot be applied in-situ when determining tensile properties. The work at hand is dedicated to expand the state of research by applying laser diffractometry (method D) to oval and elliptic fibre shapes.

Alternatively, vibroscopic methods can be used to determine the linear density. From the linear density, an average filament diameter can be calculated. Vibroscopic methods are neither proposed for carbon fibres in the aforementioned standard nor applied in the experimental section of the work at hand. Vibroscopic methods are prone to errors when used for fibres with high stiffness [155].

Notwithstanding the fact that the geometry of commodity carbon fibres can significantly differ from circular cross-sectional areas, this observation is often neglected. Literature on the true fibre shape was published (cf. Table 2.2) but when it comes to mechanical testing of fibres, it is often disregarded. This

might be due to the difficulties of an accurate in-situ determination of cross-sectional areas. Difficulties of determining cross-sectional areas of kidney-shaped carbon fibres are for instance addressed by Kowalski [156].
Typically, three procedures are used: microscopic investigations (SEM and optical microscope) either in-situ or subsequent to measurements, vibroscopic procedures (for instance when using a FAVIMAT+) or laser diffraction sensors (for instance *Dia-Stron's* tensile testing device LEX820/LDS0200). Side-views, vibroscopic procedures and laser diffraction are all based on the assumption of perfectly circular fibres. A few examples are worth discussing here. The LEX820 comes in combination with the same laser diffraction sensor as used in this work, the so-called LDS0200 by CERSA-MCI. It is in the focus of the following review.

Laser diffraction technique and sensors

In standard BS ISO-11567 [1], method D, laser diffraction is proposed to be applied for in-situ determination of the apparent fibre diameter during tensile tests.
The laser diffraction technique originates in the determination of diameters and eccentricity of metal wires [157–159]. Illustrative presentations of the operating principle are given elsewhere [160–162]. Generally speaking, the laser diffraction technique is based on determining the distance between nodes in a diffraction pattern obtained by illumination of the object with a laser. For calculation, the following equation is applied [163]:

$$d = n \cdot \lambda \cdot \sqrt{1 + \left(\frac{2 \cdot L}{\Delta Z_n}\right)^2} \tag{2.3}$$

where n is the number of fringe nodes, ΔZ_n is the distance between the nth nodes, λ the wavelength of the chosen laser beam and L the distance between sensor and fibre. According to *CERSA-MCI*, the company employs a pulsed 15 mW laser with a wavelength λ of 670 nm. The principle is based on the assumption that fine wires generate the same distinct diffraction patterns as a single-slit, also known as Fraunhofer diffraction. The validity of this assumption and different models of the diffraction technique are beyond the scope of this work, but were discussed elsewhere [164–169]. The use of an alternative principle and its derivation were provided by Martinez-Anton et al. [170]. Influences by the optical properties of the material were mentioned in [165, 169]. However, differences between metallic wires and carbon fibres were not discussed.

Application of laser diffraction sensors

Several researchers used the principles of laser diffraction to experiment with self-built versions of laser diffraction devices, commonly discussing advantages and disadvantages in detail. Krucinska and Stypka provided detailed information about their technique of determining fibre diameters [171]. The Poisson's ratio of single carbon fibres was examined with a self-built diffraction sensor, taking non-uniform fibre shapes into account. Single slit laser diffraction was also used by Chen and Harrison [69]. Comparing measurements by SEM and laser diffraction, good agreement for carbonised and untreated PAN fibres (8 - 11 µm) was found. For comparison, human hair (about 100 µm) and PET fibres (about 200 µm) were used, yielding adequate results. It was concluded that laser diffraction was a valuable tool for the determination of fibres diameters. Methods of laser diffraction for determining carbon fibre diameters were also thouroughly discussed by Li and Tietz [172]. Key factors that affect the determination of fibre diameters were summarised: debris on the fibre might cause faulty signals and diameter variations along the fibre need to be discussed. Furthermore, non-circular fibre shapes affect the measurement negatively. It was recommended to solve the latter by taking a series of measurements and thus minimising errors. Nonetheless, this implies the assumption of circular fibre shapes. Different types of fibres in the range of 5-30 µm were measured and correlated to SEM investigation. In general, results were in good agreement but the laser diffraction results lay approximately 5 % above values obtained by SEM [172]. Meretz et al. [173] contradicted the findings of Li and Tietz [172], stating that laser diffraction and SEM show a very good correlation and that no systematic error was found. However, the work deals with gold-sputtered glass-fibres, not uncoated carbon fibres as [172]. Thus, it can be concluded that results may differ significantly due to the optical properties of different fibre materials. This finding plays an important role in the experimental part of this work.

Laser diffraction was also applied by Tzeng and Chang [143] on metal-coated carbon fibres. The elliptical nature of the underlying T300 carbon fibres was considered and to minimise measuring errors, five individual laser measurements in increments of 36° were taken into account. Alternatively, the standard [1] proposes increments of 15°. The author of the work at hand applied angles of 10° [110]. All this seems justified as the decisive factors are to obtain equidistant angles and to abort the measurement after a rotation of $180° - \varphi_s$, with φ_s being the angular step. Thereby it is ensured no value is taken twice [1, 110, 143]. Bismarck et al. [174, 175] used laser diffraction and SEM on polymer coated carbon fibres. Diameters of virgin and coated fibres were determined, applying both, SEM and laser diffraction. Values obtained by laser diffraction were used for calculating the apparent interfacial shear strength in single-fibre pull-out tests. Application on carbon fibres were also discussed in [176, 177]: Perry et al. elucidated why laser diffraction underestimates the diameter of a carbon fibre by approximately 0.33 µm compared to a microscopic optical projection method. The same procedures were compared in [178] for vapour grown carbon fibres. There it was concluded that laser diffraction is more accurate but resulted in higher measured values. There is a contradiction between the assumption of a systematic error [172, 176, 177] and the absence of one [173].

In the work at hand, laser diffraction is used in the form of a closed measuring system, the LDS0200. The same device is a component of the commercially available tensile tester LEX820/LDS0200 by *Dia-Stron*. In this device, the ends of the fibre are embedded in a resin, a pretension is applied and the apparent diameter is measured once. The angular position of the fibre is thereby arbitrary. The device runs a rocking movement of $\pm 8°$ to receive an optimised laser signal. According to the manufacturer, calibration is based on stainless steel wires and causes an uncertainty of 0.01 µm.

Applications of the LDS0200 and the *Dia-Stron* testing device are found in several publications, which are outlined in the following:

Pickering et al. [179] used the diffraction sensor to determine the diameters of recycled and virgin carbon fibres. Comparative tensile tests using the cardboard method were run but no statement how the diameter was determined was given. A brief investigation of recycled and virgin fibres with the same apparatus is given by Fischer and Schmid [180]. Carbon fibres derived from a cellulose/lignin precursor were also studied with LDS0200 [181–183].

Albeit the work at hand deals with carbon fibres, it is worth taking a look to the field of natural fibres as the same scientific problem arises: how to use a testing device that gives apparent diameter readings instead of cross-sectional areas for fibres that are non-circular?

Natural fibres were investigated with the same laser diffraction sensor for instance in [184–189], without specifying if the non-circularity of naturally-grown fibres was considered. In [187–189], the diffraction sensor is not only used on single fibre level but also for bundles, not stating how interstices between the filaments were taken into account. Interstices have the same effect as rough surfaces or cavities, as the laser technique overestimates the real area. Commonly, natural fibres have cross-sectional areas less uniform than those of carbon fibres. Thus, the errors are very distinct [190] and larger than for carbon fibres. Yet, a look into the field of natural fibres should trigger a discussion about the applicability of using apparent diameters for area calculations.

With regard to the determination of the fibre diameter, most of the publications on the application of the LDS0200 for carbon and natural fibres remain vague. Details of the measurement procedure are not published and it is not stated whether additional calibrations were undertaken or if the factory-set calibra-

tion using stainless steel wires was utilised. Neither the amount of individual measurements per fibre nor the fact whether the fibre was being rotated is usually published. Alternative evaluations of the diameter readings obtained by the diffraction sensor are not recommended or discussed. It is therefore concluded that circular fibre shapes are assumed in these works, neglecting other possible fibre shapes. This evidently contradicts standard ISO-11567:2018 [1], which clearly recommends rotating the fibre within the sensor and taking a series of diameter measurements. In this context, the work at hand offers an enhancement, described in Section 3.1.
A work that stresses the importance of measuring the diameter of each individual fibre is presented by Morimoto and Ogasawara [191]. Instead of laser diffraction, another laser-based technique (laser scanner *LSM-500* by *Mitutoyo Corp.*) was used to investigate diameter variations along ceramic fibres. Measurements took place in increments of 1 mm along a distance of 100 mm. No comment if the laser scanner was rotated in respect to the fibre axis is given, but SEM was used for investigating the circularity of the fibres.
Bunsell [163] summarised that its advisable to coat translucent fibres (for instance glass, basalt) by metal deposition. Despite the fact that carbon fibres are opaque by nature, a sputtering procedure is considered in this work for verifying the applicability of the LDS0200 to carbon fibres. As the brief review reveals, the reliability of the diffraction sensor system LDS0200 is often taken for granted. Despite the widespread use of the LDS0200 in the fields of carbon and natural fibres, doubts about its applicability on these types of fibres remain and are being researched in the work at hand.

Application of vibroscopic diameter determination

FAVIMAT+ is another semi-automatic testing device frequently used for single fibre testing. It combines the determination of the diameter with subsequent mechanical testing. The diameter is determined by a vibroscopic method based on [192]. For example, the FAVIMAT+ was used for single fibre testing of carbon fibres [106, 107], for precursors and carbon fibres [193] and for cotton [105]. The FAVIMAT+ was also used to determine the linear density and the tensile strength of carbon fibres in studies concerned with the sizing chemistry [194, 195]. To determine the impact of recycling on the tensile properties, the FAVIMAT+ was used for virgin and recycled glass and carbon fibres in [108, 196]. Lyons et al. [193] discussed the possibilities of vibroscopic methods in detail.

Titze and Hunter [197] found a distinct effect upon the normal mode in vocal ligaments caused by the non-uniform cross-section. Their doubts about neglecting the true cross-sectional area of the vibrating string should also be taken into account when investigating carbon fibres as the same source of error can occur. Vibroscopic measurements are also reliant upon an accurate measurement of the density which needs to be determined prior to testing, particularly if changes in density due to recycling might be expected.

Comprehensive studies concerned with the reliability of measuring the diameter of carbon by FAVIMAT+ are not available.

Application of optical and electron microscopes

Given the vast amount of publications where fibre diameters are obtained by visual inspections (SEM, optical microscope), no detailed review can be given here. The difficulties arising when measuring the cross-sectional areas of carbon fibres during tensile tests are for instance thoroughly discussed by Jones and Duncan [33]. It was also elucidated that the rough, "crenelated" fibre surface complicates accurate measurements. Optical microscopes were correlated with side views, comparing a width (diameter) with top views measured on microsections. The non-uniformity of carbon fibres and the resulting measuring complexity is also explained in [172]. Assuming circular shapes based on SEM investigations is explicitly mentioned in [30]. Difficulties of optical evaluation of cross-sectional areas are also emphasised in [141]: optical microscopes are prone to errors because they approach the resolution limit [1];

electron microscopy is too elaborate for routine application. In addition, subsequent investigation of the primary fracture surface are complicated by the fact that the fibres tend to show multiple fractures. A remarkable publication concerned with the difference between cross-sectional area and apparent diameter was published by Thomason et al. [190]. Apparent diameters of natural fibres were microscopically determined and the corresponding circular cross-sectional areas were calculated. Compared to the corresponding cross-sectional areas found by SEM and OM, the circular calculation lead to an increase of the cross-sectional area by a factor of two. It was stated that these errors had an significant impact on determining the mechanical properties. It was concluded, that measurements relying on assuming circular fibre shapes, are not an attractive method for accurately determining the cross-sectional area of fibres [190].

Whilst SEM, microsections and microscopic observations are expedient to get an impression of the shapes of fibres within a population, they are inconvenient when it comes to large the quantities of fibres required for reliable mechanical tests.

Summarising cross-sectional areas

It was shown that carbon fibres can be assigned to four different categories in terms of fibre shape (Figure 2.7 and Table 2.2). The more complex the fibre shape, the more complex the measurement of the cross-sectional area. Common in-situ testing procedures such as vibroscopy or side-views are only applicable for circular fibres. The determination of cross-sectional areas of carbon fibres remains challenging.

Laser diffraction has become an important tool to determine the apparent diameter of carbon fibres prior to testing. Albeit it is a handy and widespread tool to determine the cross-sectional area in-situ, its applicability is not always confirmed. The work at hand evaluates to which extent oval and kidney-shaped fibres can be measured accurately by means of laser diffraction.

2.1.7 Tensile properties of carbon fibres

After having introduced procedures to acquire the shape of carbon fibres, the tensile behaviour of carbon fibres is researched in the following.

Non-linear behaviour of carbon fibres

Carbon fibres exhibit a behaviour which is described as non-linear, non-linear elastic or non-Hookean. This non-Hookean behaviour is also referred to as a stiffening of the fibre or strain stiffening. Despite the different designations, it is one and the same effect: the higher the strain region, the higher the corresponding Young's modulus. The phenomenon is caused by changes of the microstructure and is observed in single fibres as well as in composites. In the following, a brief review of literature is presented.
Regarding single fibres and fibre strands, several observations have been published, with the first description of the effect being released by Curtis et al. in 1968 [2]. The phenomenon of non-linearity was investigated by ultrasound methods in HT and HM fibres. In the first third of the load range, rapid changes of the dynamic Young's modulus in relation to the load were observed, but above this threshold, the increase of the modulus responded linearly to the load. Furthermore, it was stated that no hysteresis occured and that the effect was reversible. It was concluded that the non-Hookean effects originated from reorientations of crystallites in the graphitic fibre structure [2]. Derived from this work, two different Young's moduli can be defined and are used within the work at hand: the resting Young's modulus E_0 is defined as the extrapolation of individual calculations of elasticity to zero strain. In contrast to this, the Young's modulus as determined at any other point is labelled E_a, the apparent modulus, in accordance with Curtis et al. [2].
Hughes [141], who has investigated single fibres and impregnated strands, recommended the following relationship to describe the non-Hookean effect:

$$E_\varepsilon = E_0 \left(1 + f \cdot \varepsilon\right) \tag{2.4}$$

with E_0 being the resting modulus (zero strain), E_ε the modulus at strain ε and f a factor depending on the type of fibre. In this context, E_ε and E_a are the same variable. Kowalski [156] introduced an empirical model, adding a second-order term to Hooke's law and verified the findings for single fibres, fibre strands and composites. There, the secant modulus determined at strains between 0.1 and 0.6 % are essentially the same for composites and strands but slightly reduced when testing single fibres. Hughes [141] emphasised the importance of stating the strain region in which basic material properties were determined, a fact that is still often neglected today. Djordjević [198] observed non-Hookean elastic behaviour for four types of PAN-based HT and HM carbon fibres under investigation. In addition to a linear function as mentioned by Hughes [141], a second-order polynomial regression was used to deduce the resting modulus. It was found that both approaches yield approximately the same value. Nonetheless, they recommended second-order polynomial fitting [198] and both are investigated in the present work. Beetz and Budd [199] investigated pitch- and PAN-based carbon fibres by applying a strain modulation technique. It was found that different moduli responses occured for fibre made from different precursors, concluding that the microstructure of the fibre affected the non-Hookean behaviour. Moreover, a novel fibre gripping apparatus was used for the measurements to reduce the effects of the system compliance.

Guigon et al. [20] did not describe non-Hookean behaviour in detail, but defined an initial, a secant and a final modulus, with the latter being almost 20 % higher than the initial (resting) modulus. No clarification why three moduli were used to describe the tensile behaviour instead of observing a continuous stiffening was given. The findings should be interpreted as non-Hookean behaviour. Twelve different HT carbon fibres were investigated in TEM using dark-field and selected-area electron diffraction (SAD or SAED) techniques. By observations of the microstructure, a close packaging index, which correlated linearly with the Young's modulus and particularly the tensile strength was defined. Guigon et al. did not compare the microstructure in unloaded and loaded state and thus could not directly correlate the non-Hookean behaviour to changes in the microstucture. A relation between non-Hookean behaviour and microstructure was presented by Huang and Young [200]. A narrowing of the Raman peak was observed during loading of the fibres and could be correlated to an increase of the orientation of crystallites in the fibre. Arsenovic et al. [201, 202] examined non-linear behaviour by means of laser-generated ultrasound. It was stated that a better alignment of crystallites resulted in an increased initial modulus but also that the remaining potential gain in modulus is less pronounced when the fibre is then strained mechanically.

Using X-ray diffraction Shioya and Takaku [203] also observed an increased crystallite orientation when tensile stress was applied to carbon fibres. In a later work, Shioya et al. [204] created a mosaic model, which was parabolic in nature and fitted experimental results well. With increasing tensile strain, the deformation of crystallites was more and more constraint. Investigating the tensile properties of heat-treated carbon fibres, in high resolution transmission electron microscopy (HRTEM) and SAD, Zhou et al. [36] proposed a mechanistic model of the stiffening effect: under tension, misorientated crystallites were gradually aligned with the tensile direction. Entangled junctions between two crystallites first slid and then locked, which lead to an increased load transfer between adjacent crystallites, paired with an increase of the tensile modulus. On the basis of this review, the non-Hookean behaviour of carbon fibres can be derived from the structure-property-relationship and changes undergone due to tensile loads.

Based on publications in this field [141, 200], stress and Young's modulus of a carbon fibres are plotted as a function of strain in Figure 2.8. The higher the strain region, the higher the modulus of the fibre. Extrapolating several calculations of Young's modulus E_ε, determined at defined strains ε, yields the resting Young's modulus E_0.

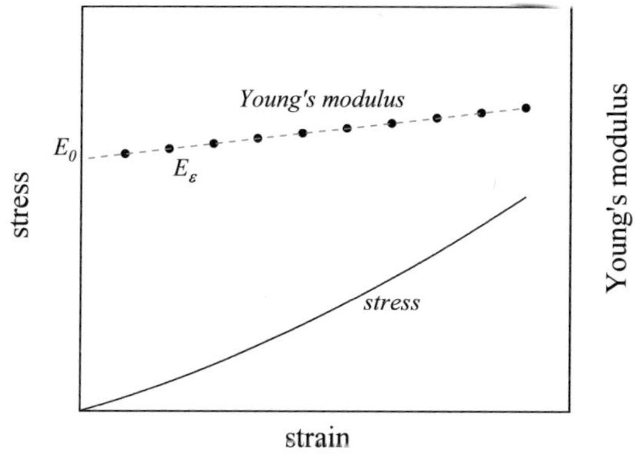

Figure 2.8: Theoretical stress-strain and Young's modulus-strain diagram of carbon fibres with non-Hookean behaviour based on literature [141, 200].

In terms of conventional single fibre testing, standards recommended compliance corrections to account for contributions of the load train and the interaction of fibre and gripping system [111]. Compliance corrections should be made with the same type of fibres, gripping system and testing conditions as present in the subsequent testing [111]. As Kant and Penumadu emphasised, this procedure is prone to errors for fibres that exhibit non-Hookean behaviour themselves [116].

The non-linear mechanical behaviour of carbon fibres is also addressed in the work at hand: direct strain measurements as presented by Reder at al. [125] and by the author [109, 110] allow for more detailed investigations of the non-Hookean behaviour as effects caused by the system compliance do not interfere with the experiment.

Regarding carbon fibre reinforced composites, the non-Hookean behaviour was extensively studied [205–211] since the first discovery by Hughes [2] and the most important observations are summarised here. The non-Hookean behaviour was observed in tensile and also in compressive mode [206, 212]. Summarising literature, an increase in the Young's modulus of approximately

20 - 30 % can be expected when comparing strain regions close to 0 % and at approximately 1 % under tensile load [205–208]. The effect is particularly pronounced in unidirectional composites, as described in [205, 209, 211, 212]. Bogdanovic et al. [205] stated that the effect was less pronounced for ±45-laminates and Lagace [209] added, that the non-linear stiffening effect was higher when the fibres were aligned in testing direction. If there was no alignment along the fibre direction, neither rearrangements of the crystallites nor the associated increase in stiffness would occur. Instead, softening of the composite occurred due to shear [209]. Furthermore, Lagace [209] stated that only composites loaded approximately in the load direction showed no hysteresis effect. If the angle between load and fibre direction exceeded 20°, loading and unloading cycle did not follow the same path.

In summary, carbon fibres and their composites tend to exhibit non-Hookean behaviour. Thus, it is expedient to always state in which region of the stress-strain-curve and with which method the Young's modulus was determined. However, studying data sheets of several suppliers, this is often not the case. Interestingly, Ishikawa et al. [207] observed that the Young's modulus provided by their supplier corresponded exactly to the average modulus between zero stress up to failure, not the resting modulus.

Failure behaviour of carbon fibres

Not only the fibre's non-Hookean behaviour before failure but also the failure itself has been the subject of many studies. Failure of carbon fibres is believed to be described well by theories based on Griffith's models [213, 214].
Summarised in [24], surface and internal flaws potentially initiated carbon fibre failure. An investigation of Johnson and Thorne [65] demonstrated that by surface etching, artificial flaws could be introduced. Etching changed the effect of the original flaws. Higher tensile strengths were measured after the etching, offering potential for the modification of carbon fibres. Additional literature

concerning the strength-structure relationship and models of fibre failure is for instance found in [38, 215, 216]. The presence of misorientated crystallites is found to be an important factor to initiate tensile failure [215, 216].
The assumption that the two types of defects, namely surface flaws and internal flaws, exist, allows for applying Griffith's and Weibull's theories to describe fibre failure. Thus, the failure behaviour of carbon fibres is commonly described by Weibull statistical distribution function [217], albeit in the year of Weibull's work, the era of carbon fibres was just at the dawn. Among other examples, Weibull himself attributed the fibre strength of Indian cotton to his distribution function [217]. Weibull's statistics were applied in several publications to evaluate the strength of carbon fibres, for instance in [137, 218, 219]. In [93], the applicability of Weibull's theory to carbon fibre was confirmed, but it was discussed that amendments were needed to apply it for dry or impregnated fibre bundles.
Besides the widespread application of Weibull's theory to evaluate single fibre tests, doubts remain about its applicability: Thomason [220] stated that Weibull's theory can only be used if the whole range of fibre strengths is included in the experimental data. He argued that fibres too weak to be handled are discarded before testing and thus the results are shifted to higher strength values [220]. The undesirable effect of unintentionally selecting stronger fibres was also described by Chi et al. [221]. Concluding, Thomason recommended large quantities of fibres to be tested, with a computer model yielding 80 fibres as threshold [220]. Zok [222] communicated distinctions between weakest-link-theory and Weibull's statistics and general misconceptions about the two. He summarised that Weibull's statistics can be applied as long as not more than one flaw population is confirmed or a model indicates that it would not be applicable. Despite few exceptions, it is common practice to use Weibull's theory to evaluate the strength of brittle carbon fibres.

Tensile properties of carbon fibres found in literature

In this section, tensile properties of carbon fibres presented in literature are collected and reviewed. The Young's moduli and tensile strengths of carbon fibres in relation to their cross-sectional area and the gauge length are presented. Note that some publications do not inform whether properties were determined in bundle or single fibre tests.

Tensile properties of carbon fibres and their relation to fibre diameter

The work at hand aims at revealing the structure-property relationship of single fibres and composites. An overview of characteristic mechanical values from various literature sources about PAN based carbon fibres is presented here.

There is general consensus that a relation between fibre diameter and the mechanical properties exist. For illustration purposes, several works are outlined visually in Figures 2.9 and 2.10: the larger the fibre diameter, the lower the tensile strength and the Young's modulus appear to be. Once more it has to be noted, that "diameter" is a term misleadingly used in literature, indicating that circular fibre shapes are assumed. In addition, values are commonly rounded to represent a large fibre population and do not consider single fibres as in the work at hand. Most of the sources do not provide information about the origin of the data, for instance, how the diameter was determined and what kind of mechanical testing procedure was used. In addition, fibre treatments are also commonly neglected in publications.

In case values varied between two different sources, both are presented as variations might be due to the year of production or the testing method. Nonetheless, the collection of data from several sources provides a broad insight into the relationship between geometry and properties. Here, all fibres with moduli beyond 310 GPa are defined as HM fibres, all others as HT or IM fibres. The threshold varies slightly from the 300 GPa recommended by IUPAC [61]. Even the definitions by IUPAC are ambiguous, as the range for IM fibres overlaps with the ranges of Young's moduli of HT and HM. High modulus fibres

are displayed as open symbols, HT and IM as solid symbols. Symbols with the same geometry correspond to the same data source, which are Peebles et al. [223], Fitzer et al. [58] and Chung [24].
From data sheets[6] only HT CF and IM CF data were collected and considered in Figures 2.9 to 2.11. The bold black × represents the value from the data sheet of PANEX 35, the reference fibre in the work at hand [4].
Where needed, the diameters are transferred to circular cross-sectional areas via $A_{CSA} = \pi\, r_{fibre}^2$ to account for the evaluation of non-uniform shapes as discussed later. What has been said applies equally to Figures 2.9 to 2.12.
For both types of fibres, reduced cross-sectional areas correspond to higher values of the Young's modulus (cf. Figure 2.9). The scatter is higher for HM fibres (with a Pearson correlation coefficient 0.44) than for the cluster of HT and IM fibres (Pearson correlation coefficient 0.89). In addition to the graphical representation of the effect, some more works are worth mentioning: The correlation of a decrease in Young's modulus due to an increase of the cross-sectional area was also summarised by Kobets and Deev [146], who further related this to changes in density. De Lamotte and Perry [224] showed a strong correlation of decreases in tensile strengths and Young's moduli of HT fibres diameter of approximately 8 - 12 µm. It was inferred that microstructure variation are among the reasons for these changes and that controlling the diameter of precursor fibres is essential to produce high quality fibres. Strain rate dependencies, which were also investigated in their work, were neither found for tensile strength nor modulus. In an additional technical note, the correlation between mechanical properties and diameters for HT fibres was confirmed [177]. For fibres that had not undergone a final heat treatment,

[6] Data sheets available online, 04.06.2019 for:
Hexcel carbon fibres type AS4, AS4C, AS7, IM2A, IM2C, IM6, IM7, IM8, IM9, IM10 (https://www.hexcel.com/Resources/DataSheets/Carbon-Fiber); for Toray carbon fibres types T700S, T800S, T1000G (https://www.toraycma.com /page.php?id=661).
Available on request for Toho Teijin/Toho Tenax carbon fibres type HTS40, UTS50, HTS40 (https://www.teijincarbon.com/de/) and for Cytek Thornel carbon fibres type T300 and T650-35 (https://www.solvay.com/en/brands/thornel)

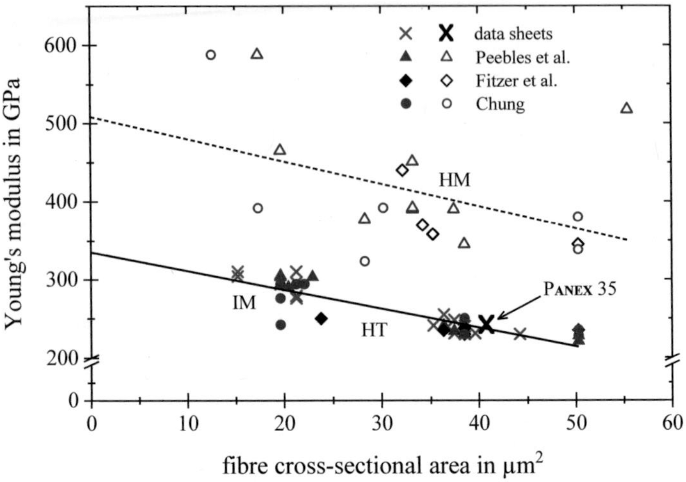

Figure 2.9: Collection of data from literature to demonstrate the correlation between cross-sectional area of carbon fibres and their Young's modulus. Data were collected in various data sheets of different manufacturers and in the literature of Peebles et al. [223], Fitzer et al. [58] and Chung [24]. Solid symbols represent data of HT and IM fibres; open symbols display HM fibres. The dashed and the black lines are linear fits of both data sets. Regions of HT and IM overlap. Refer to footnote 6 for information concerning the data sheets.

no such correlation was found. Summarising, the correlation between diameter and mechanical properties originate from the heat treatment [177]. The importance of reducing the diameter during production to obtain high tensile strength carbon fibres was discussed in [70].

Jones and Duncan [33] resumed that both, the Young's modulus and the tensile strength, increased for smaller fibre diameters due to changes in the ratio of skin to core. It was concluded that cracks in the core region were more likely misaligned with regard to the fibre direction, showing higher potential to initiate fibre failure. Cracks in the skin region were, even at the same size, less prone to cause failure as they aligned with the fibre direction.

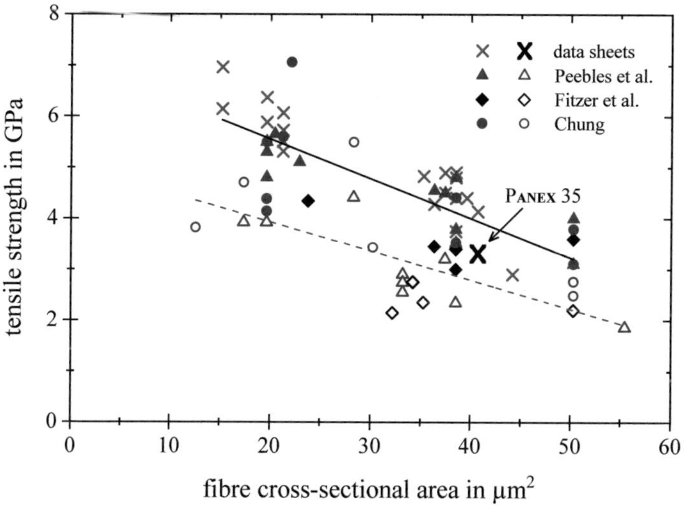

Figure 2.10: Tensile strength of carbon fibres as a function of the cross-sectional areas. For all types of fibres displayed here, tensile strength is decreased for increased average diameters [24, 58, 223].

The correlation between the tensile strength and the cross-sectional area of several types of fibres is illustrated in Figure 2.10. In this representation, the clamping length and thus the fibre volume cannot be taken into consideration due to missing data. Thus, possible effects due to variation in fibre length according to the weakest-link-theory cannot be evaluated here.

Commonly, the Young's modulus of carbon fibres is plotted versus tensile strength, for instance in [8, 9, 225]. It has been reported that tensile strength and Young's modulus are opposing physical quantities in carbon fibres [7, 9]. Plotting the data taken from different sources, presented in Figure 2.11, emphasises this finding.

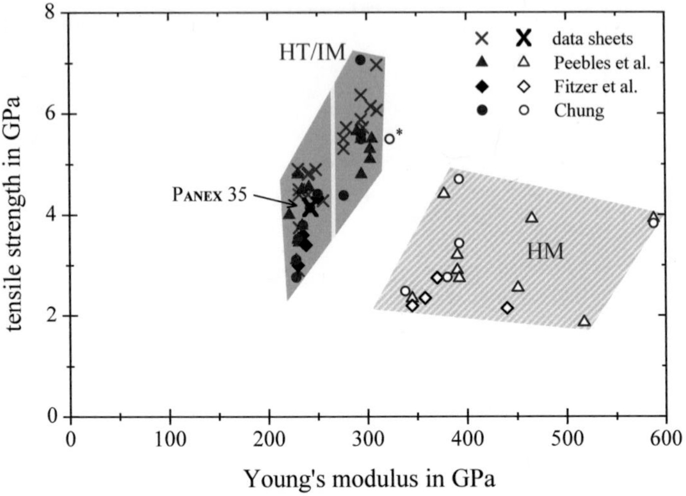

Figure 2.11: Tensile strength plotted versus the Young's modulus: two clusters appear and HM fibres tend to have decreased values of tensile strength. /* was excluded from the clustering.

Two clusters appear in Figure 2.11; one of HT and IM, the other of HM fibres. HT and IM could be further subdivided into two separate clusters in the range of approximately 240 GPa (HT) and 300 GPa (IM) [58]. Fibres with intermediate modulus present a compromise between HT and HM fibres and research in this field is ongoing [9]. It is worth resuming that reduced cross-sectional areas relate to an increase in mechanical properties [7].

Tensile strength of carbon fibres as a function of gauge length

The so-called "size-effect" is a well-known and well-described phenomenon in tensile testing of fibres [102, 226]. As variations in fibre diameter are often small, it is frequently simplified as a correlation between gauge length and tensile strength. The size effect describes the observation, that a decrease of the gauge length commonly results in an increase of the tensile strength. The effect is assigned to the weakest-link-theory: the larger the volume of a specimen, the more likely the occurrence of a larger flaw from which a crack is initiated when the specimen is loaded. Some publications concerned with this effect are summarised and covered below. Again, the description of the effect originates in the field of cotton testing, for instance [227]. More recently, Pan et al. [228] investigated polymer fibres (PP, PET, PA) and carbon fibres. For all, a significant and continuous reduction in tensile strength and breaking strain was found when increasing the gauge length from 10, over 20 and 50 to 100 mm. Running Weibull analysis on two different sets of pitch-based carbon fibres, Chwastiak et al. [229] demonstrated the correlation of higher tensile strengths for shorter gauge length. For instance, both types of fibres showed an increase of more than 50 % for a gauge length of 1.4 mm compared to 40 mm. A decrease in tensile strength for larger gauge lengths was also published by Diefendorf and Tokarsky [31] for high modulus fibres and by Watanabe et al. [230] and Naito et al. [137] for high strength fibres. Deng et al. [231] investigated sized and unsized/desized carbon fibres. While the sizing had only a minor influence, increasing the gauge length caused a reduction in tensile strength for all fibres. Gauge length effects were also investigated by Moreton [232]: several differently heat-treated carbon fibres showed decreases of about 30 % when increasing the gauge length from 10 to 50 mm.

The publications mentioned above deal primarily with the influence of the clamping length and do not consider effects due to variations of the cross-sectional areas of the fibres any further. Some of the results discussed above are summarised graphically in Figure 2.12.

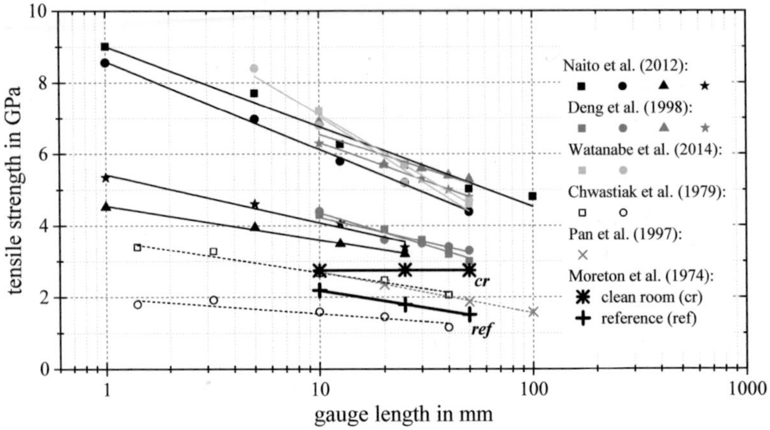

Figure 2.12: Tensile strength as a function of gauge length, as found in publications: Naito et al. [137], Deng et al. [231], Watanabe et al. [230], Chwastiak et al. [229] and Pan et al. [228]. For various types and years of production, the same trend is seen: larger gauge lengths correspond to decreased tensile strength as highlighted by trend lines. While the specifications of those fibres are not of interest here, special attention shall be drawn to the experiments by Watt and Moreton [64]. Fibres produced in a clean room are the only fibres which do not follow the described trend thanks to their reduced flaw density.

Moreton and Watt published a fundamental work in the field of size effects in carbon fibres [64]: producing fibres under clean room condition, it was found that the strength of these fibres was not affected by changes in gauge length (10 to 50 mm). In contrast, reference fibres, produced under normal conditions, showed a dependency on gauge length. After heat treatment, the reference fibres showed a significant reduction of tensile strength compared to the clean room fibres. Furthermore, the Young's modulus was only marginally affected by the spinning conditions, and it was concluded that the Young's modulus is an intrinsic property of carbon fibres while the tensile failure is dominated by flaws [64]. Thus, the Young's modulus should not vary with respect to the gauge length; an assumption that is dealt with in the experimental part of this work. Since other sources do not evaluate variations in the Young's modulus and since no direct strain measurements have been reported until recently [109, 110, 117, 127], the evaluation of the literature considered in this section was limited to the tensile strength.

Poisson's ratio of carbon fibres

In light of the fact that it is challenging to determine the cross-sectional area of carbon fibres and their real strain during tensile tests, it is demanding to determine the Poisson's ratio of carbon fibres. Only few approaches are reported in literature: Krucinska and Stypka [171] introduced an experimental method using laser diffraction to determine the Poission's ratio. Values between 0.26 and 0.28 were obtained with an accuracy of 5 to 7 %. Villeneuve et al. [233] studied thermal expansion and the Poisson's ratio of ceramic and carbon fibres by scanning electron microscopy. For Poisson's ratio, values of only 0.1 to 0.22 were reported and the experimental difficulties faced during the experiments were addressed. In a review, a range of 0.025 to 0.3 were stated for the Poisson's ratio of carbon fibres but no further information regarding the experimental background was provided [234]. This large variation emphasises the need for novel experimental procedures to determine the Poisson's ratio. Nowadays, the Poisson's ratio is of increasing interest to feed data into computational models created for improved predictions of composite properties.

Concluding remarks on carbon fibres

Resulting from the state of research as reviewed so far, a few statements shall be given:

- The mechanical properties of carbon fibres are highly dependent on the production method and fibre structure. For instance, fibre diameters are a consequence of the production parameters and influence the tensile properties.

- Many technical and natural fibres do not have perfect circular shapes. Therefore, it is worth finding approximations of the real cross-sectional area instead of dealing with apparent diameters based the on assumption of circular cross-sectional areas.
- Prior to further investigations, for instance mechanical testing, the cross-sectional area of each individual fibre needs to be measured to guarantee accurate measurements. This is the only way to minimise measuring errors caused by the assumption that fibre bundles contain uniform fibres.
- As fibre diameters may vary along the fibre axis, it is advisable to either take several measurements along the fibre or to investigate the cross-sectional area in the position where failure occurred. For obvious reason, the latter is only possible if plastic deformation during testing can be excluded. Moreover, carbon fibres tend to fail at multiple positions due to their brittle behaviour. It is difficult to reclaim the fragments for additional investigations.
- Characteristics of the fibres must be taken into account: the non-Hookean behaviour of carbon fibres impedes the correction of the system compliance of the test rig. It would be advisable to state the strain region which was used for determining the Young's modulus.
- Direct strain measurements, which are common practice for macro-scale specimens, are also recommendable on fibre level when the Young's modulus and strain to failure are investigated.
- In order to ensure statistical reliability, the testing of a large quantity of fibres is recommended. Semi-automatic systems can assist in this task.

2.2 Carbon fibre reinforced polymers and their interface

Starting from the comprehensive discussion of carbon fibres in general, their utilisation in carbon fibre reinforced polymers (CFRP) is now at the centre of attention. It starts on composite level, introducing carbon fibre reinforced polymers, their production via the bulk moulding compound route and recycling of carbon fibres. This is accompanied by a brief reflection on interface properties, a description of the chemical nature of the carbon fibre surface and how to gain knowledge of the surface properties of carbon fibres.

2.2.1 Introduction to carbon fibre reinforced polymers

Composites are defined as materials consisting of two or more individual phases that are homogeneous on macroscopic level but heterogeneous on microscopic scale. The components can contribute in different aspects, for instance mechanical properties, costs, conductivity etc. For a composite to be successful and efficient, the properties to be optimised should differ by a factor of three or more in the individual components [235]. Exemplary, the Young's moduli of typical thermosets are approximately 3 GPa whereas the reinforcing carbon fibres can easily exceed a value of 300 GPa. Concerning the composite structure, the fibres carry the loads, while the matrix is responsible for holding the fibres in place, determines fibre alignment and enables load transfer [226, 235]. In material science, particularly in the description of composites, four so-called paradoxes are defined [226]:

- **Paradox of real strength**
  The real strength of solid materials is significantly lower than the strength predicted by calculations [236].

- **Paradox of fibre shape**
  Fibres show significantly higher tensile strengths than the same material in any other form. The thinner a fibre, the higher its tensile strength [213].

- **Paradox of fibre length**
  The longer the section of fibre under observation, the more likely a large defect exists in the fibre and the lower is the measured tensile strength. As a consequence, the shorter the distance between the clamps in a fibre tensile test, the higher the tensile strength of the fibre [226].

- **Paradox of composites**
  Composite materials containing aligned fibres have the potential to withstand higher loads than its individual components would [237].

All four paradoxes are premised on imperfections in the microstructure [226]. Especially the paradox upon gauge length plays an important role in the work at hand when discussing single fibre strength.

The world of composite is extensive, with manifold composite components (fibres, fillers, etc.), reinforcement materials (carbon fibre, glass fibre, ceramic particles etc.) and matrix materials (polymers, metals, ceramics). Equally diverse are the type of fibre length (short fibres, long fibres etc.), fibre orientation and architecture (unidirectional, biaxial etc.) and processing technique (resin transfer moulding, bulk moulding compounds, injection moulding etc.). To keep this introduction short, only bulk and sheet moulding compounds are reviewed in this section. For additional information, the reader may refer to [80, 226, 238]. In addition to the selected production routes, the interface properties are discussed and recycling in the context of carbon fibre reinforced composites is reviewed.

2.2.2 Sheet and bulk moulding compounds

Bulk moulding compounds (BMC)[7] and sheet moulding compounds (SMC) are interesting materials when it comes to lightweight design for the automotive sector [239]. Bulk moulding compounds and sheet moulding compounds are akin in their production routes and application, as both rank among quasi-isotropic composite materials containing randomly orientated fibres [80]. Commonly, long fibres (1 to 50 mm) are used as reinforcement.

In this section, similarities and distinctions between BMC and SMC are elucidated. Capabilities of recycling via a BMC route are discussed in preceding sections. A brief description of both process routes is given in accordance with Figures 2.13 and 2.14. For historical reasons, both materials typically consist of glass fibres and unsaturated polyester resins (UP), which are often supplemented by fillers and additives. Together, SMC and BMC accounted for one quarter of 2018's production in the sector of glass-fibre reinforced polymers [240]. Still, carbon fibres have a minor market share, accounting for less than 2 % of reinforced polymers [240]. The share of carbon fibres in the BMC/SMC market might grow in the years to come, if recycling is further promoted. Recycled and loose carbon fibres might compare to virgin glass fibres in regard of their costs and might become even more attractive to use [241].

The process routes of BMC and SMC

Both process routes start with the provision of the resin paste. The resin components, additives and fillers are mixed and prepared for further processing steps. The desired fibre volume content is usually converted to mass contents and the correct amounts of resin paste and fibres are dosed gravimetrically. However, the procedures to combine fibres and resin differ greatly between

[7] In literature, the terms "dough moulding compounds" (DMC) and "bulk moulding compounds" (BMC) are used to describe the same material. It is henceforth referred to as BMC, as this is nowadays the common term in Germany.

SMC and BMC process routes [80]: in SMC technology, Figure 2.13, the resin paste is filled into two reservoirs in equal amounts. From the reservoirs, resin paste is dispensed as a thin layer onto a polymeric carrier film by the help of a doctor blade. The film passes a chopping unit and is covered with fibre bundles cut to the desired length. A second carrier film with resin paste is then placed on it bottom up. This sandwich passes a compactor unit consisting of several rollers. The material is coiled up and prepared for maturing and subsequent part production.

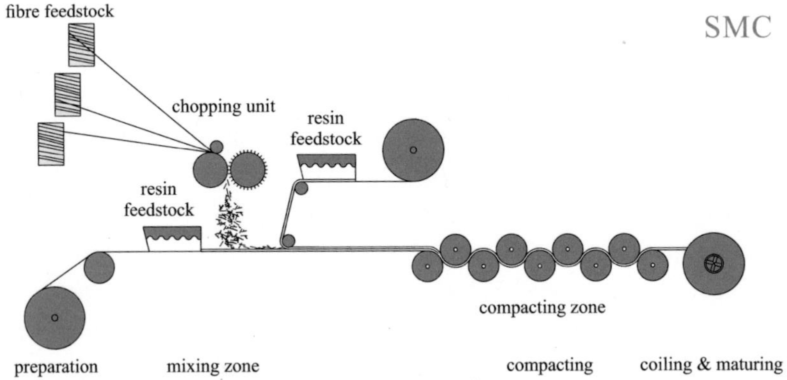

Figure 2.13: Graphical representation of the sheet moulding compound (SMC) process; *partly based on* [80].

The main step of the discontinuous BMC process, Figure 2.14, is the blending of the resin paste with cut fibre rovings in Z-kneaders until all fibres are thoroughly impregnated. When mixing is completed, the bulk mass is removed from the kneader and sealed gastight in polymer bags in the desired amount. These semi-finished products are then matured and stored until further processing takes place.

For both, SMC and BMC, the sealed packages are matured based on a two-step chemical reaction without fully curing the material. In the final step, BMC and SMC can be processed by compression moulding [96]. For BMC, injection moulding is an alternative [242]. Curing takes place in this step and being thermoset materials, no further remodelling is possible thereafter. As for instance shown in [150], SMC and BMC semi-finished products can be moulded into the same sheet-like product.

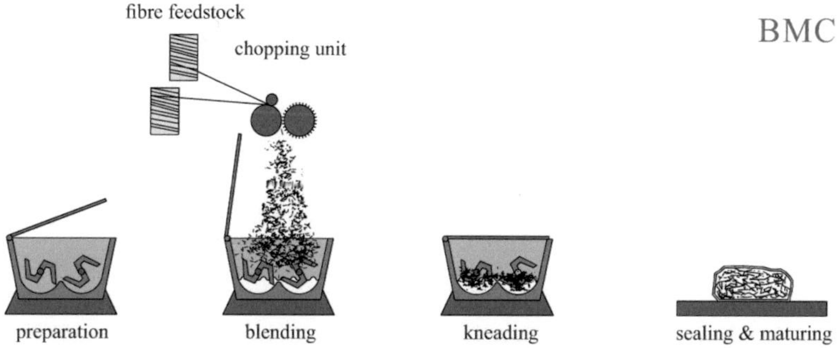

Figure 2.14: Graphical representation of the bulk moulding compound (BMC) process. Images of the production steps are shown in the experimental part, Figure 4.4.

As the name SMC suggest, the semi-finished product is sheet-like and commonly, sheet- or troughlike products, for instance bootlids and cover panels, are produced by the SMC route [96]. The usefulness of mass production of SMC parts for automotive applications was already recognised in the 1980's, then based on glass fibres [243]. It was realised that SMC and BMC were cost-effective alternative in lightweight design [244]. BMC routes can be applied to produce components such as casings and headlight reflectors. The maturing and curing cycles are dependent on the specific chemistry of the resin.

Summarising, the provision of the base materials and the maturing and curing steps are very similar for SMC and BMC materials. Process-wise, the main difference is the way fibres and resin are conflated. Structure-wise, fibre bundles in SMC materials are supposed to lie flat in the plane, whereas some bundles can be orientated perpendicular to it in BMC materials. Both structures contain fibre bundles rather than random clusters of single fibres [150]. It is worth mentioning that typical BMC materials contain less than 30 wt.-%

fibres [80]. Notwithstanding this fact, it was feasible to obtain a standard of 55 wt.-% fibres for the material prepared in the work at hand.

Bulk and sheet moulding compounds can compete with steel and aluminium in several lightweight applications. Compared to unidirectionally reinforced composites, they are inferior in terms of mechanical properties, but superior in terms of production costs. SMC materials have become more popular and are more frequently reviewed in literature than BMC. However, comparing SMC and BMC processes (Figures 2.13 and 2.14) one can see that the machinery needed is less complex for BMC. Furthermore, a homogenisation occurs after the blending step. Resin, fibres and optionally fillers are thoroughly kneaded. In SMC, an even application of the fibrous reinforcement needs to be ensured as only minor compaction and homogenisation take place afterwards. Particularly, continuous semi-finished fibre products are needed as feedstock. In this respect, the selection of added materials is more versatile for BMC. Hence, BMC technology might prove advantageous in the coming years when it comes to recycling.

The following advantages and disadvantages, summarised in [80, 96], apply equally for BMC and SMC:
Both are cost-effective, particularly in small and medium-sized serial production, as tool costs are lower than for steel sheet moulding. Manifold geometries can be realised, even those that would be challenging to mould with metals, for instance ribs. Lightweight design is possible, with composite densities of approximately 1.5 to 1.8 $g\,cm^{-3}$ [245]. The heat resistance up to approximately 190 °C is sufficient for most automotive applications. A major drawback are low tensile strengths, caused by the random orientation of the fibre bundles. What was said applies analogously for glass and carbon fibre reinforcements for SMC/BMC [80, 96].

2.2.3 Carbon fibre textiles – wovens and non-crimp fabrics

The production of carbon fibres was detailed in Section 2.1.3. Commonly, the fibre rovings are subsequently wound on bobbins. From there, they can be further processed as any other textile: they can be knit to wovens, tacked to non-wovens, stitched to non-crimp fabrics or weaved to three-dimensional geometries [246]. Alternatively, fibres can be cut to be used in the SMC or BMC process, presented in Figures 2.13 and 2.14. Comprehensive information on fibre fabrics can be found in [235, 246].

In wovens, fibre rovings undulate and decrease the mechanical properties of the composite. Non-crimp fabrics eliminate this issue and thus gained industrial importance [247, 248]. They are often used as stacks to produce dry preforms, which are then impregnated and moulded. Preparing dry preform stacks, up to 40 % of the material end up as dry fibre fabric waste, rendering these process routes material-intensive [249]. Woven and non-crimp fabrics are no standard materials when it comes to the feedstock of standard BMC processing (cf. Figure 2.14). They can, however, be a valuable resource for recycled CFRP's [150, 250, 251].

2.2.4 The fibre/matrix-interface in carbon fibre composites

The interface plays an important role in fibre-reinforced polymers. It is created by the solidifying step during composite production, either due to curing (thermoset) or cooling (thermoplastic). To benefit from the concept of reinforcement, a defined load transfer from the polymer matrix into the fibre has to be guaranteed. If the load transfer is not optimised, solely positive locking instead of a chemical bonding is present.

Resumed in [252], composites can be studied on four levels: on molecular level, interaction between matrix and fibre occurs on basis of chemical and physical bonding, for instance by chemical bonds and van der Waals forces. The second level is the micro level. Interfacial action between fibre and matrix are investigated and for example described by interfacial shear stress, bond

strength and critical energy release rate. On meso level, the distribution and orientation of fibres within a matrix is considered. The macro level is needed to characterise the properties of specimens and structural components.
To correlate composite properties with interface properties, the meso level is of interest. Geometric approximations of the fibre packing in composites are shown in Figure 2.15. For uni-directionally reinforced composites with dense packing structure, rectangular and hexagonal arrays were proposed [234, 253], presented in Figure 2.15 a) and b), respectively. In part c), fibre contours found in the micrograph of a real BMC material are shown. It is evident that neither the real fibre shape nor the packing is as uniform as in the models given in a) and b).

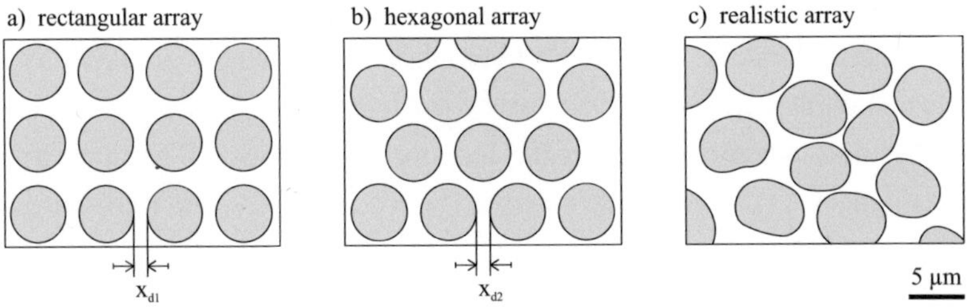

Figure 2.15: Packing of fibres in composites: a) rectangular and b) hexagonal packing, which are both theoretical structures. c) shows a realistic packing structure, with neither fibre shape, position nor distances x_{d1} and x_{d2} being uniform; *a, b based on* [234]

The interface itself is typically comprised of the molecular and the micro level. However, micro and molecular level are usually discussed mutually [252, 254]. A comprehensive overview of the interface in carbon fibre/epoxy composites is presented by Hughes [234]. Some summarising statements are: a certain disorder of the fibre surface increases interfacial bonding. Surface oxidation can be beneficial for HT carbon fibres, whereas positive locking is expedient for HM fibres. Sizing and surface treatment can increase the adhesion; however, a positive effect is dependent on the combination of fibre and matrix.

When fibre rovings are used as feedstock for BMC processing, fibre bundles are apparent in the composite product. Process-related, these fibre bundles are randomly orientated. A certain amount of bundles is aligned rather transversely to the loading direction and such regions are prone to failure. Hughes [234] distils four possible failure mechanisms seen in transversely loaded composites, as shown in Figure 2.16. Exemplary shown are fibres (grey) in a polymer matrix (white). The figure illustrates the step from meso to micro level.

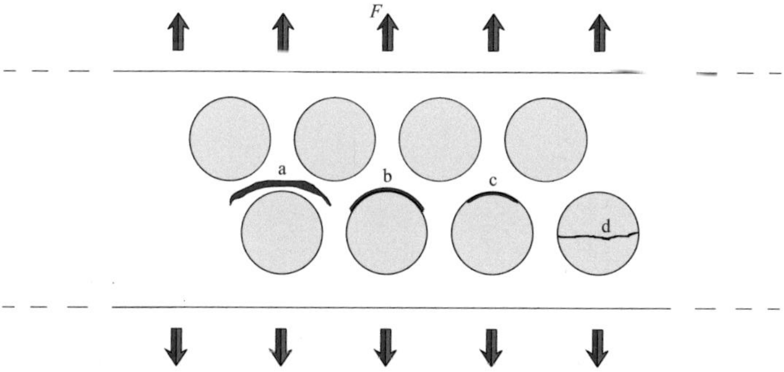

Figure 2.16: Failure mechanisms in transversely loaded fibre reinforced composites. Cracks can occur in the matrix (a), the interface (b) or split the fibre close to the surface (c) or through the core (d); *based on* [234]

If the fibre is stronger than the matrix and a good interface exists, failure is initiated in the matrix (a), in a region where fibres are packed densely. A weak interface causes failure by debonding of matrix and fibre at the interface (b). In case the fibre has weak traverse tensile properties, it might split close to the surface (c) or through its core (d). Fracture type (d) is rarely encountered in carbon fibre composites [234]. Scanning electron microscopy of fracture surfaces can assist in defining the failure mechanism, albeit small traces of resin on the fibre might be difficult to detect. Partial debonding can be desired in terms of absorbing energy during crack opening [234].

In terms of testing of the interface, the most popular techniques were evaluated by Zhandarov and Mäder [252]: interface testing can be divided into two groups: the first consists of experiments where the external force acts directly on the fibre, as in push-out [255, 256], pull-out [257] and microbond tests [258, 259]. These are schematically shown in Figure 2.17 and all investigations in the work at hand focus on this group. Alternatively, the fibre can be embedded in a matrix, with load being transferred from the matrix to the fibre. Examples are the single fibre fragmentation test and the Broutman test [252].

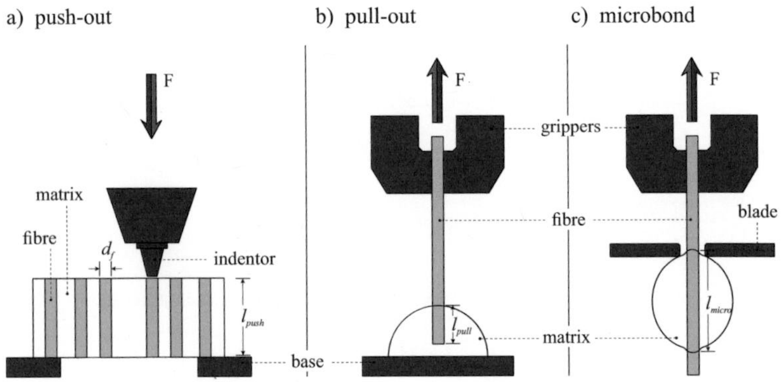

Figure 2.17: Concepts of testing procedures to determine the fibre/matrix-interface according to [252].

An important interface property, the interfacial shear stress τ, is calculated by Equation 2.5 in accordance with [260]:

$$\tau = \frac{F_{max}}{A_e} = \frac{F_{max}}{\pi \cdot d_{fibre} \cdot l_e} \tag{2.5}$$

with F_{max} being the maximum load, d_{fibre} the diameter of the fibre under investigation and l_e the length of the embedded fibre. Analogous equations are used to determine the interfacial shear stress in pull-out or microbond tests. The embedded length l_e depends on the testing procedure. For the push-out test (a),

it is the thickness l_{push} of the polished specimen, for the pull-out (b) it corresponds to the depth l_{pull} to which the fibre is embedded in the matrix and for the microbond test (c) it is the length l_{micro} of the droplet.
The influence of fibre alignment in respect to the push-out direction has been modelled and studied by Brylka et al. [261, 262] for reinforced thermoplastics. Effects on push-out results due to of fibre misalignment are important for BMC materials. Given their randomly-distributed fibre bundles, it is impossible to prepare push-out samples with perfectly aligned fibres. Brylka et al. expected a force increase of 21 % if a fibre was misaligned by 25°. However, their computer model only calculated a deviation of 3 %, presumably due to earlier damage [262]. Due to the conclusions stated in a critical review [263] of the microbond test, it is not executed in the work at hand: the microdroplet has an influence due to variations in shape, the reproducibility of its preparation and its response to mechanical loads. Consequently, loads might be introduced and distributed non-uniformly.
Interface properties are particularly challenging in terms of fibre recycling. Recycled fibres have likely lost their sizing due to the harsh conditions during the recycling of the composite, for instance when pyrolysis or solvolysis are used. Here, the molecular level can be influenced significantly. Dry fibre waste still contains sizing. However, the origin of the fibre waste is commonly unknown and thus all information of sizing content and type has to be gained experimentally if an optimised interface is demanded.

2.2.5 Characterisation of the physical-chemical fibre properties

The work at hand aims at a comprehensive understanding of the impact of recycling on carbon fibres. To bridge the gap between single fibre properties and effects on the composite, the fibre surface was investigated. It is expected that the chemistry of the fibre surface is affected by the recycling process and that the sizing will be gone. In this section, a short overview how to gain knowledge of the surface properties is given. More specifically, dynamic vapour sorption (DVS) and inverse gas chromatography (iGC) are introduced briefly. Both methods aim on revealing differences between virgin and recycled carbon fibres.

Inverse gas chromatography is a versatile and sensitive tool to investigate physicochemical properties of materials. It has become a widespread standard procedure for many applications and details are found elsewhere, for instance in the review [264] or in a work concerned with technical fibres [265]. It is commonly combined with BET[8] [266] measurements, for instance to investigate recycled carbon fibres [267]. Application of iGC to carbon fibres are, for example shown in [268, 269].

DVS has evolved in the field of drug and food technology [270, 271]. The technique is used to investigate the relationship between a solid and sorbed vapour, typically water vapour [270]. Dynamic vapour sorption determines the change in mass of a solid sample as a function of time while the sample is exposed to changes in the surrounding vapour concentration. In case water vapour is used, one refers to the term "relative humidity" to define the vapour concentration [270]. Whilst the field of application of DVS has extpanded to natural fibres [272], studies with carbon fibres have not been reported in literature yet.

2.3 Recycling of carbon fibres and their composites

Firstly, this section begins with a brief evaluation of incentives to recycle and it examines the complexity of recycling in terms of carbon fibre reinforced composites. Secondly, possible routes of CFRP and carbon fibre recycling are introduced. Thirdly, a closer look at the mechanical properties on single fibre level is taken.

[8] Brunauer-Emmett-Teller theory

In today's world, it is doubtless that society, economy and research need to have sustainability in mind. One possible contribution to sustainability lies in recycling and recycling-fair construction [273]. In general, principles of recycling-fair construction have been known for decades [274]. Recycling itself is by no means a recent invention as re-using materials has been along tradition in human history. Nonetheless, since the 1970's, the awareness of the need for sustainability and recycling has grown [273].

Introduction to recycling of CFRP

In the automotive sector, the demand for lightweight vehicles rises in order to meet the ambitious goals of high mileage claimed by regulations [275]. Yet, a growth of the composite market is accomponied by an increase in the volume of parts to be recycled. Landfill will no longer be an option [276] and thus strategies for the recycling of vehicles, including CFRP parts, are needed [277]. Since 2015, 85 % of a vehicle are required to be recycled and since 2007 it is the responsibility of the manufacturers to take care of the disposal of end-of-life vehicles [277].

The world of recycling is diverse. Technologies for various material classes are available or are being researched, but it would go far beyond the scope of this work to discuss them all. Recycling procedures differ significantly between composites and fusible, homogeneous materials such as glass or aluminium [241].

Recycling technologies show high complexity [10]. Thus, whenever new materials and processes to produce components and goods are developed, specialised recycling technologies will lag behind, usually for several years. As recyclability is commonly not considered a major endeavour in industrial production [274] and as political and societal pressure had been relatively low, new materials can enter markets before their reusability and sustainable waste management are completely clarified [278]. In particular, this applies to carbon fibre composites, which gradually increased market share even though market-ready, industrial-scale recycling procedures were not available [241]. To date, only pyrolysis of CFRP and mechanical grinding of GFRP show a high technology-readiness-level [279].

For the recycling of composites, a dilemma arises, as two opposing goals are present. Engineers steadily increased the properties of composites. This goes hand in hand with an optimisation of the interface and increasing the cohesion between fibre and matrix [234]. By contrast, to recycle composites, fibre and matrix have to be separated. Obviously, these objectives are contradictory. Reclamation is commonly in favour of the more expensive fibres, refraining from regaining the low-cost matrix [278]. Costs of several base material of composites are presented in Figure 2.18 [245, 280].

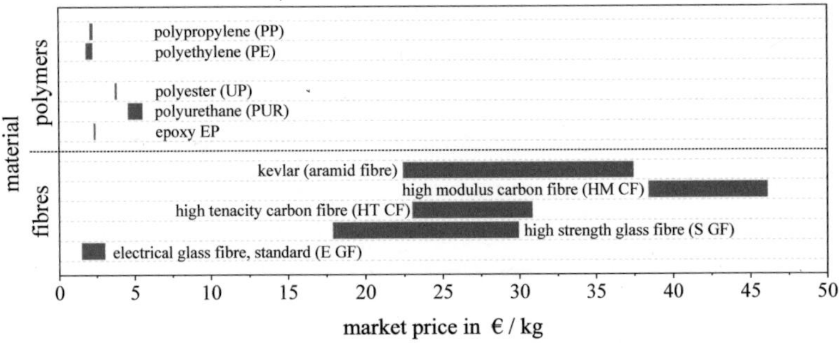

Figure 2.18: Costs of base materials used in composites: except for standard E-glass fibres, all fibres cost significantly more than polymers [245, 280]. According to [281], recycled carbon fibres are expected to cost 12 to 17 € per kg.

Costs per kilogram for virgin carbon fibres outweigh those of common polymers and standard glass fibre by an order of magnitude. It is economically attractive to recycle carbon fibres while using the matrix material for energy recovery. This is commonly achieved by pyrolysis [10]. Alternatively, solvolysis is being researched to recover both, fibre and matrix [10]. The market price of recycled carbon fibres is expected to range between 12 to 17 € per kg [281]. The potential added value is far higher for carbon fibres than for glass fibres.

Challenges of composite recycling are summarised in [278]: composites are itself a mix of at least two component (matrix and reinforcement) but are often enriched with additives, for instance fillers. Applying hybridisation concepts, composites are frequently combined with different material groups. Examples are metal inserts [282], foams [283] and elastomers [284]. Hybridisation complicates the separation of the components even further and recycling of hybrid materials is addressed in more detail in [285, 286]. Separation and classification can be costly and difficult. The supply chain is less stable than for traditional materials such as aluminium and steel [278]. Waste management needs to take care of possible releases of gaseous, liquid and/or toxic by-products. Some aspects of the potential hazards of recycling are addressed in more detail in [287].
Several boundary conditions of interest were presented by Yang et al. [285]: a) the availability of waste and the economic aspects of large scale operations, b) the quality of the recyclate and whether they are compatible with the existing composite market, c) environmental aspects, d) overall costs and profitability of recycling technologies and e) the availability of technologies. The last-named boundary condition is discussed in the following.

Material flows in CFRP recycling

Composite waste does not only arise in form of end-of-life parts. On the contrary, within different production steps, different types of waste are generated. Thus, one has to clarify if the focus is upon the recycling and disintegration of a composite, mainly at the end-of-life, or upon the reuse of production waste, particularly non-impregnated fibre waste (dry fibre waste). In the work at hand, both are considered as recycling.

Recycling of reinforced polymers was extensively reviewed by Pimenta and Pinho [10, 84], Oliveux et al. [281] and by Asmatulu et al. [278]. Based on these investigations, the flow chart in Figure 2.19 was created. The upper section shows different stages of the composite material over the passage of time. In the lower section, the different waste fractions that arise are emphasised. There, the potential recycling routes are graphed. The adjacent numbers refer to examples from literature described in the text. Up to 40 % of the composite waste are expected to arise as production waste, that is dry fibres and infiltrated spoilt [226, 249].

Within the first step, raw materials such as polymers, fillers and fibres may be excluded due to poor quality, weighing errors, exceeding of storage periods etc. Among this fraction are fibre fabrics, like woven or roving coils. This fraction (R-I) plays an important role in the work at hand. It is considered "dry" waste, as it has not been in contact with a polymer matrix. Its chemical and mechanical properties are assumed to be the same as in the virgin material. Large amounts of carbon fibre fabric waste occur during the draping of dry preforms for the manufacturing of complex structures, as can be seen, for example in the studies of [248]. Summarised in [10], 40 % of CFRP waste occurs during manufacturing, where again 60 % is dry fibre fabric waste.

Another relevant waste fraction are prepregs (R-II), as they are prone to material degradation and thus storage periods need to be satisfied strictly. Compared to dry fibre waste, prepregs are more complex to be recycled due to pre-cured polymer matrix already adhering to the fibre [10].

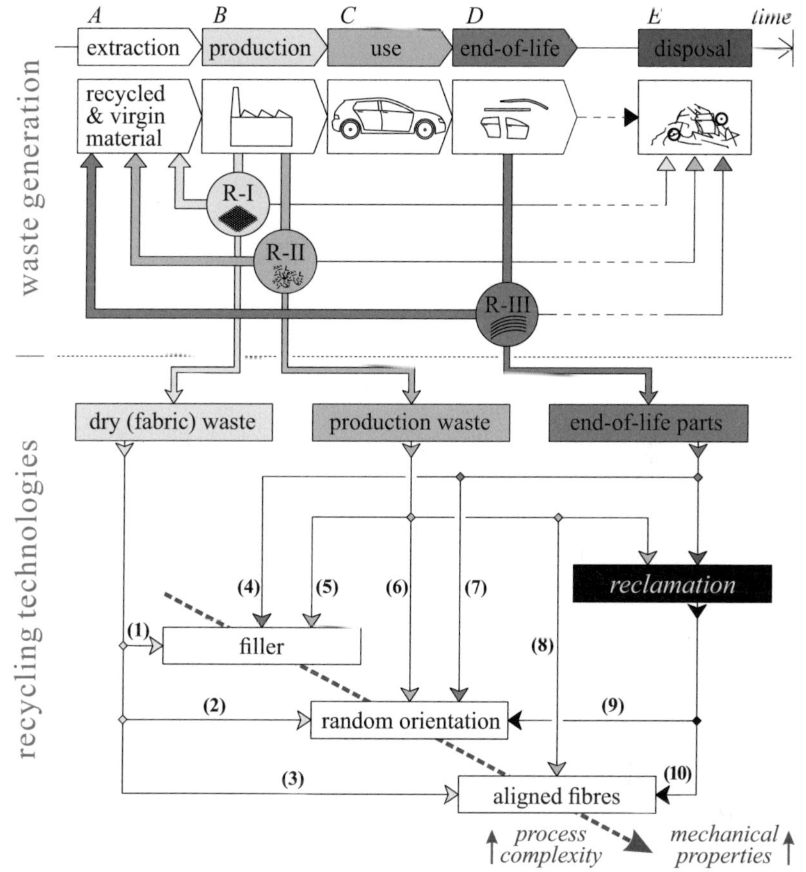

Figure 2.19: Flow chart of state of the research in composite recycling. Top: waste generation during different production steps of fibre reinforced composites, for instance cars. Bottom: routes of recycling technologies and fibre reclamation. Number (1) to (10) represent recycling routes that have been researched. Examples for each are presented in the text. *Partly based on* [10, 84, 278, 281]

After infiltration and cooling of the thermoplastic or curing of the thermoset, respectively, spoilt can result from production faults such as wrongly calibrated machinery, design faults or material defects. As production of complex structures may require cutting and trimming steps to achieve the desired contour, composite waste is generated. At last, components turn into waste after exceeding their life span and the so-called end-of-life parts need to be recycled or deposited. As Figure 2.19 depicts, all waste materials might end up on landfills, although this contradicts the demand for sustainability. Up to the present day, composite recycling focusses on the reuse of the fibres, as their value, particularly that of carbon fibres, is more than a magnitude higher than the value of standard polymers used in composites. Instead of disposing of the composites as waste, recycling strategies are needed. This is the focus of the lower part of Figure 2.19. Dry fibre waste, R-I, is the most promising waste as its original properties are maintained. Thus, they can re-enter the production cycle, either ground as fillers (1), or with random orientation (2), or after an intermediate step to achieve a certain fibre alignment (3). If resin and fibres are successfully separated, the fibres can be re-used as reinforcement analogously to the re-use of dry fibre waste (9, 10). Nowadays, pyrolysis and solvolysis are the most promising reclamation techniques [10]. Using reclaimed fibres as fillers is not presented in literature, as it would be ineffective to use highly complex processing routes to gain a filler. The reclamation step is not always necessary. Instead, production waste (R-II) and end-of-life parts (R-III) can be shred and ground to be used as fillers (4, 5). Alternative, ideas to reuse both types of wastes directly in composites with random orientation (6, 7) have been investigated. In what follows, approaches for each of the recycling routes are given. This corresponds to a selection without the claim to completeness. Here, the focus lies upon the feasibility of recycling routes rather than the mechanical properties.

Generally speaking, recycling of remnants and production waste in-house is favourable as this waste is customarily not contaminated and its exact properties and characteristics are known. Thus, reusing this waste within the primary production route is often possible. Obviously, this is equally true for glass, plastic and dry fibre waste.

(1) The application of milled carbon fibres as conductive filler in a conductive graphite-epoxy composite material was investigated [288]. An increase in electric conductivity compared to the reference was found. Carbon fibre fillers produced with recycled materials are also available commercially[9].

(2) Procedures for a direct BMC process re-using dry fibre fabric waste were introduced by the author in [150, 251]. It is the composite material used in the work at hand.

(3) Hybrid yarns were investigated in [289]. Recycled dry carbon fibre waste was simulated by cutting rovings into lengths of approximately 40 and 60 mm. Thereby it was feasible to research the influence of fibre volume content and fibre length of the carbon fibre "waste" combined with polyamide 6 to manufacture hybrid yarns. Yarn tenacity increased with increasing rCF fibre length and volume content.

(4/5) To produce filler materials from prepregs, they need to be cured. They can then be ground to fillers in the same way as composite waste or end-of-life parts. End-of-life parts were ground and used as filler in SMC and BMC processes [290]. CFRP was crushed in [291] to produce reinforced, injection moulded ABS specimens. Tensile stress and stiffness were increased with higher ratios of rCFRP filling: tensile strength reached a plateau for 50 wt.-% filler, tensile stiffness was increased up to 70 wt.-% of recycled filler. Examples for using ground composites as fillers in injection moulding were also presented by [292, 293]. Ground composite waste was found to be a beneficial filler in highly-filled composites, typically SMC and BMC, as it reduces the density of the composite when replacing the standard filler calcium carbonate [294].

[9] For instance: https://www.car-fibertec.de/cf-flex - 07.05.2019

(6) In [295], prepreg waste of different origin was cut and screened. It was used to manufacture BMC products with a fibre volume content below 10 % and partially aligned SMC. The re-use of prepreg waste by cutting it into small pieces and then remoulding it was discussed in [296, 297].

(7) Reinforcements by randomly orientated pieces of CRFP without an intermediate step of fibre-resin-separation were investigated in [298], where CFRP scrap was used to reinforce concrete. The average size of the CFRP was up to about 21 × 8 mm. Work of fracture was increased even for large pieces, allowing for a satisfiable reinforcement without the need to reclaim individual fibres.

(8) Prepreg waste can be aligned [295] by processes based on paper-making technology. The same discarded prepreg material as in (6) was used to manufacture rCFRP SMC mats with fibre contents up to 44 %. Tested in tensile mode, a distinct anisotropy between machine direction and transversal direction for the aligned rCFRP products was found. It was superior compared to materials from virgin glass fibres which were assumed to have the same costs as the recycled prepreg material [295]. In another work, production waste was sorted by using a hammer mill, a classifier and sieves to obtain four different types of recyclates. The most favourable, a fraction consisting of fibre bundles with lengths up to 10 mm and only minor amounts of resin residue was used to manufacture SMC [299].

(9) Aerospace carbon fibre waste was reclaimed by pyrolysis and used for injection moulding of polyphenylene sulfide [300].

(10) An interesting approach in closed-loop recycling was made at the Universtiy of Bristol [301]: there, a process named HiPerDiF (High Performance Discontinuous Fibre) was invented. Starting point was a suspension of carbon fibres in water. The suspension was accelerated towards a nozzle and hitting a conveyor belt, thereby aligning the fibres in a

preferred direction. This preform was impregnated, moulded and then mechanically characterised. By means of pyrolysis the fibres were reclaimed and the procedure started from the beginning. In theory, this would lead to an eternal and closed-loop recycling and results after one recycling cycle were promising. However, the authors had to face a decrease of stiffness and tensile strength after the second recycling run. Nonetheless, the procedure is promising in terms of recycling of end-of-life parts. The need of fibre alignment to obtain rCFRP with increased mechanical properties was also emphasised in [302].

These literature reviews revealed that three basic concepts of fibre recycling can be followed. After reclamation, the fibres can be treated and conditioned, for instance to guarantee high-quality surfaces. In order to design high-performance recycled composites, elaborate procedures to align the fibres are needed. The opposing concept forgoes these costly procedures: a reduced mechanical performance is accepted in favour of lower costs and a straightforward processing [150, 251]. A particularly simple process variant is to mix the recycled reinforcement with virgin matrix material and produce parts by injection moulding. This concept is being pushed to the extreme, when the fibres are not even reclaimed but instead, ground composites are fed directly into the injection moulding machine [292, 293].
Summarising the material perspective, the reclaimed fibres can be reused as fillers, as randomly orientated reinforcement or as re-aligned reinforcement.

Tensile properties of recycled carbon fibres

To this point, this review focussed on fibre reclamation and recycling procedures presented in literature. In the following, the focus is shifted to tensile properties of recycled fibres.
Albeit the work at hand deals with carbon fibres as reinforcing materials, it is worth giving a statement concerning glass fibre recycling. There, higher losses in tensile strength due to recycling were observed: regarding Weibull tensile

strength, Pickering et al. [303] found decreases of 50, 80 and 90 % for process temperatures of 450, 550 and 650 °C, respectively. Fibres were reclaimed in a fluidised bed process. In the context of the work at hand, another factor is interesting. As no stiffness loss of the fibres was observed, BMC casings were produced, mixing virgin and recycled fibres in a 1:1-ratio, demonstrating the applicability of BMC for recycling. Reduction of fibre strengths up to 70 % were also found in [304]. It appears that glass fibres are more affected by the temperatures needed to dispose of the resin [305]. However, deeper insights exceed the focus of this work and the reader may refer to the appropriate literature for more detailed information concerning glass fibre recycling. The comparably large loss in tensile strength of recycled glass fibres makes carbon fibre recycling even more attractive.

During the last decade, several works on the recycling impact on carbon fibres have been published. Some of them are briefly summarised in the following. Hereby, the focus lies on the tensile properties of single fibres to underline the importance of single fibre testing. The recycling strategies to reclaim the fibre can be found in the accompanying publications. Again, deeper insight is found elsewhere [10, 281, 306].

Pimenta and Pinho [84] dedicated themselves to the comparison of composites based on virgin and recycled fibres. Their study considered single fibres and re-manufactured composites with woven carbon fabrics. Fibres were pyrolysed at a commercial facility, *ELG Carbon Fibre Ltd*, with four different parameter sets. Whilst milder conditions (lower temperatures) lead to unchanged fibre diameters and area densities of the fabric compared to the reference, harsh recycling condition (500 - 700 °C) reduced the values by approximately 20 and 40 %, respectively. In terms of single fibre tensile strength, evaluated with Weibull statistics, only the mild conditions yielded fibres as good as the reference. For all other recycling parameters, fibre strength dropped to 33 % or less of the reference values. Investigating fibres in SEM, it was found that harsher conditions led to increased surface flaws, so-called pitting. Whilst pitting was not found for the mildest conditions, almost 8 % of residual resin remained on the fibres, rendering their use in subsequent composite production more complicated. The tensile strength of the single fibres was also reflected in the composite properties: the harsher the conditions of the pyrolysis, the greater the loss in composite tensile properties. Tensile stiffness remained between 64 and 95 % of the reference; tensile strength only reached 26 to 68 % compared to the reference.

In addition, major challenges of the recycling process itself were emphasised: the origin of the carbon fibre waste is commonly unknown and the recycling process time is often inversely proportional to the remaining fibre quality. Particularly, processes need to be transferred from batch processing to large quantity, continuous processes [84].

In two studies, Piñero-Hernanz et al. [307, 308] recycled carbon fibres by hydrolysis in near- and supercritical water and by solvolysis in sub- and supercritical alcohols. Hydrolysis led to recycled fibres with 85 to 98 % of the tensile strength of the virgin reference [307]. For the alcoholic solvolysis [308], single fibre tensile tests confirmed that 85 to 99 % of the reference tensile strength were retained, depending of the selected parameters. Minor resin residue was observed in both routes. The amounts of resin residue depended on the recycling conditions.

Jiang et al. [309] used a chemical procedure to reclaim carbon fibres from an epoxy composite. Single fibre tensile tests revealed that 95 % of the tensile strength and the original fibre stiffness compared to virgin fibres were retained. With only about 20 fibres being tested for each state and not applying a Weibull procedure, statistical assurance does not seem to be guaranteed. Another chemical solvolysis approach was presented by Liu et al. [310]. Decomposition temperature, nitric acid solution concentration and the weight ratio of sample to the chemical solvent were the parameters under investigation. Single fibre tensile tests were run with a rather high testing speed of $10\,\mathrm{mm\,s^{-1}}$. With an optimised parameter set, the strength loss was 1 %. In the worst case, the strength was decreased by 13 %.

Fernandez et al. [311] investigated the pyrolysis of carbon fibres at a temperature of 500 °C and subsequent oxidation. Tensile strength was lowered by approximately 10, 20 and 30 % when increasing the oxidation time from 30 min to 60 min and 90 min. The oxidation phase was used to remove char residues. The apparent diameter was lowered up to 12 % compared to virgin material.
Nahil and Williams [267] reclaimed carbon fibres by pyrolysis and added a subsequent oxidisation. Pyrolysis was run at five different temperatures from 350 to 700 °C. Oxidisation took place at 500 or 700 °C. For oxidisation temperatures of 500 °C, the retaining tensile strength was 67 to 93 %. For oxidisation at 700 °C, only 26 to 36 % of the virgin fibre strength were retained. At least 50 single fibres were tested for each parameter; tensile strength was defined as the average value obtained.
Greco et al. [312] also introduced a post-treatment of the fibres subsequent to pyrolysis. Compared to virgin fibres, a loss of tensile strength up 70 % was observed. However, only values of rCF obtained in their laboratory with vCF properties stated in the data sheet were compared. Weibull analysis was performed, albeit only 15 individual fibres for each parameter set were evaluated, an amount too low if Thomason's arguments are considered [220]. In terms of diameter determination, the work remained unclear and results obtained via SEM do not seem to be reliable.
Stoeffler et al. [300] investigated single fibres and polyphenylene sulfide composites. Tested on single fibre level at a gauge length of 10 mm and a test speed of 1 mm/ min, recycled fibres showed a reduction of 22 % in terms of tensile strength compared to virgin fibres. Composites were injection moulded using in-house produced pellets, some enriched with recycled, others with virgin fibres. In terms of tensile strength and modulus, consistent or better tensile properties were found for recycled material compared to the reference containing virgin fibres. This was likely due to the fact that the fibres were not the same: the recycled composites contained a mix of waste from aerospace industry whilst the virgin reference was PANEX 35, a rather low cost commodity

fibre. Material-wise, this approach might have been inconclusive as the virgin and recycled fibres were not of the same source and type. From a holistic and economic point of view, high-grade carbon fibre waste can serve as reinforcement and compete with virgin materials, particularly when a straightforward process such as injection moulding can be used, as demonstrated in [300].

The studies presented here, for instance [84, 267, 307, 308], show that the tensile strength of recycled carbon fibres can be deteriorated significantly compared to their virgin counterpart. The extent to which tensile strength is reduced depends on the recycling procedure. Therefore, it is not possible to put forward universally valid theses. Each individual case must be considered, and possible effects due to the recycling need to be studied. Comparability is also complicated by different single fibre testing procedures. As elucidated in detail in Sections 2.1.4, the standards currently available for single fibre testing do not guarantee reproducibility from one facility to the next. It is noticeable that publications concerned with the recycling of fibres do not elaborate single fibre testing procedures in detail. Most of them use standard cardboard frame methods without specifying the whether system compliance was evaluated. It must be assumed, that the determination of the cross-sectional area of the fibre was not optimised for non-circular fibres and that direct strain measurements were not used. Particularly, the former is interesting in the field of recycling as a reduction of the cross-sectional area due to recycling conditions must be feared. In most of the publications presented in this section, values for the Young's modulus were stated. However, as neither an evaluation of the system compliance, nor procedures of direct strain measurements were presented, a detailed comparison does not appear to make sense here. In a nutshell, only minor effects of fibre recycling on the Young's modulus were reported. In the work at hand, enhanced testing procedures are developed and shall be applied for both, virgin and recycled carbon fibres.

In [313], it was argued that ecological benefits of CFRP recycling were only given if the recycled materials can be used to replace virgin fibres. In closing, some alternative shall be mentioned: carbon fibre reinforced polymers contain mainly carbon as element. Shredded CFRP can readily be used for incineration thanks to its high calorific value. The heat can then be used for heating or the generation of electricity [226]. Alternatively, ground CFRP can be used as filler in construction materials [281].

An apt summary of composite recycling was presented by Woidasky [241]. Recycling of CFRP shows higher potential than recycling of GFRP as the fibre material itself is significantly more valuable. CFRP recycling becomes most attractive when a high fibre alignment can either be reclaimed or can be restored cost-effectively.

Difficulties of comparing composites manufactured with virgin and recycled fibres were addressed in [306]. The difficulty is to achieve a fair and viable comparison between both. On single fibre level, this can be readily achieved by keeping preparation and testing conditions constant. On composite level, the production process and its parameters are reflected in the mechanical properties. If virgin and recycled CFRPs are to be compared, all parameters must be adjusted. For instance, fibre length, type of fabric and production route need to be the same. From a scientific point of view, it makes, for instance, no sense to compare virgin fibres processed via dry fabric preforming with composites based on non-woven recycled fibres. The direct BMC process used in this work [150, 251] works well in this context, as both types of fibres can be fed directly into the BMC kneader and all subsequent processing steps (maturing, pressing, curing) are the same.

Ultimately, however, one of the most prominent boundary conditions of recycling is only partly a technical issue. Rather, the question is whether industry gains confidence in using recycled materials in composite structures or if it avoids taking any risks [281]. This completes the state of research of this work and the resulting scientific questions are presented.

2.4 Scientific question

As revealed in the preceding sections, recycling of composites and its subcomponents is gaining importance. In order to create a market for recycled carbon fibres and to use them for future design of structures, creating a profound knowledge of possible influences by the recycling process is inevitable. Particularly, the single fibre properties are of interest.
Thus, the following research questions are deduced and are at the centre of attention in this doctoral thesis:

- How can single fibre properties be evaluated with higher accuracy than proposed by standards and publications?
  Due to the importance of this question, an entire chapter (Chapter 3) is dedicated to this subject.

- To what extent are carbon fibres affected by recycling and does this lead to a significant impact on the composite properties?
  This question is evaluated from microscopic to macroscopic scale by various experiments, described in Chapters 4 to 7.

This work focusses on material investigations rather than on recycling techniques. To describe and understand the properties of composites, it is advantageous to understand the fibres, the matrix and the interface and, most of all, their interactions. In addition, the surfaces of carbon fibres and their chemical properties are investigated.

According to the scientific relevance, the scope of the work is on single fibre testing and single fibre properties, but always keeping an eye on interface, composites and the interaction between all three levels. Where possible, the materials are examined using procedures given by the state of research in material testing. However, in some field, the state of research had to be further developed to understand material properties. This applies in particular to the testing of single fibres in terms of the Young's modulus and the determination of the cross-sectional areas of irregularly shaped carbon fibres. In this field, the work at hand gives suggestions how to overcome inaccuracies caused by indirect strain measurements and the erroneous assumption of circular fibre shapes.

3 Development of a single fibre testing method

The overview of procedures for testing single fibres presented in the state of research emphasised the need of further optimisation of measurement setups and techniques. Thus, this entire chapter deals with possible enhancements of single fibre testing procedures. Starting point is the definition of requirements and the design of the experimental setup. Subsequently, novel procedures to determine strain and cross-sectional areas are presented and discussed together with respective literature. Focus and verification lie upon carbon fibres but are presumably applicable likewise to other types of fibres.

3.1 Design and method development

For the test setup to be developed, several requirements were defined and shall be listed here. While some are inevitable to evaluate the effect of recycling on single fibre properties, others are optional and offer additional advantages for future applications of single fibre testing in general applications. The following mandatory requirements were defined:

- **Strain measurement:** The design of the testing device was supposed to offer an solution how to enable direct strain measurement. Direct strain measurement refers to tracking the elongation of the fibre instead of recording the movement of the crosshead of the testing device. This avoids disadvantages of the system compliance as it renders all correction calculations redundant.

- **Cross-sectional area:** A way to approximate the non-circular shape of carbon fibres was sought in order to obtain more reliable results of the cross-sectional areas. Improving procedures to determine fibre strain and the cross-sectional area are directly related to improvements in determining the Young's modulus by equation 2.1.
- **Fibre alignment:** Another factor to increase the accuracy in single fibre testing is the correct alignment of the fibre in tensile direction.
- **Fibre length:** The device was supposed to cover a wide range of fibre lengths and to enable a stepless adjustment of the distance between the grippers. This offers flexibility, for instance the clamping length can be adjusted to recycled fibres, which are commonly truncated during the recycling processes.

Optional adjustments to serve the goal of advancing research in the field of single fibre testing were:

- **Transferability:** In the long term, a transfer of the principle into industrial applications should be conceivable. Here, a combination with semi-automated fibre handling is desirable to reduce testing time, and consequently increasing the statistical certainty by running a larger set of tests.
- **Modularity:** By using individual modules, the device can be adapted to the specifications of different fibres and different testing modes. The module for measuring the cross-sectional area can be replaced to adapt to transparent fibres. The load cell can be adjusted to different demands. Particularly, the fingers can be re-designed to cope with different fibre materials.
- **Versatility:** The device was developed for testing carbon fibres, but is capable to working with other types of technical and natural fibres.

A device to fulfil these goals and conditions was developed by the author and presented in [109, 110]. In the following sections, the setup and its functions are described in detail. Procedures to measure strains directly on the fibre and to determine the cross-sectional areas of elliptical carbon fibres are introduced. Subsequent to the general development of methods, the specific test implementation is presented and results obtained are discussed in succeeding chapters.

3.1.1 Design and geometry

In the following, the device designed within the frame of this work is referred to as SiFiT (**Si**ngle **Fi**bre **T**esting device) and the setup as well as its function are described in reference to Figure 3.1. There, the SiFiT is highlighted in white. It is installed within a universal testing device *Zwick//Roell 2.5 kN* (C). Two computers (A) are used for controlling the testing device and the camera (B) as well as for logging the measurements of the cross-sectional area. An ultraviolet light source (D) is needed to illuminate so-called tracking beads. A control panel (E) is used to open and close the grippers, to apply suction for fibre alignment and to adjust the air pressure used. The maximum air pressure to be applied is 6 bar, commonly it was reduced to 3 bar. The lower part of Figure 3.1 highlights the components of the SiFiT. Connectors (a) are used to install the SiFiT within the universal testing device and connect it to the crosshead and the baseplate. The load cell (b) is exchangeable and here a 5 N load cell by *Zwick//Roell* (accuracy class 0.5) was used. Connected to the load cell is the upper gripping system (c), consisting of a parallel gripper type PGN+40 by SCHUNK GmbH & Co. KG and a in-house designed pair of steel fingers. Analogous, the lower gripping system (e) was designed. Here, a PGB 64 by SCHUNK GmbH & Co. KG was used to allow for the installation of a tubule. The latter can optionally be used to create underinflation by a Venturi nozzle (f) in order to align the fibre (x). The steel fingers are coated with a layer of rubber tape to protect the fragile fibres. A laser diffraction sensor LDS0200 by CERSA-MCI (d) is used to determine individual diameter reading as data base for the elliptical fitting procedure described in Section 3.1.2.

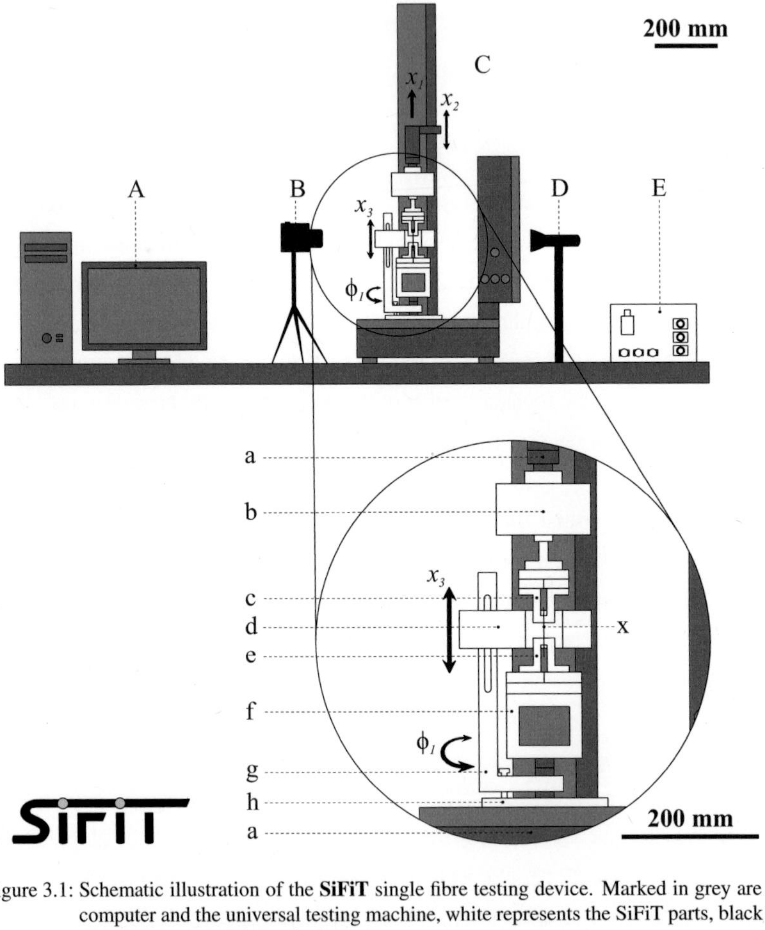

Figure 3.1: Schematic illustration of the **SiFiT** single fibre testing device. Marked in grey are the computer and the universal testing machine, white represents the SiFiT parts, black are the components of the optical system.
A) computer, B) camera for DIC, C) universal testing device, D) UV light source, E) gripper control unit; a) connectors, b) load cell, c) upper gripper, d) laser diffraction sensor, e) lower gripper, f) Venturi nozzle, g) support pole, h) perforated plate,
Labels x_1, x_2, x_3 and ϕ_1 represent the degrees of freedom.

The laser diffraction sensor sits on a support pole (g), and can be rotated around the longitudinal axis. A rotationally symmetric steel plate (h) is connected to the testing device. Bore-holes are drilled in angles of 10° and a pin is used to fasten the support pole (g) in different angles. Angular steps of 10° were chosen based on the concepts presented in literature [1, 143].
The arrows in Figure 3.1 are used to highlight the test direction and the degrees of freedom of the setup. The testing direction is given by (x_1). By moving the crossbar prior to testing (x_2), the clamping length of the fibres can be adjusted. Independently, the laser diffraction sensor (d) is height-adjustable (x_3) with the standard height setting being the lengthwise centre of the fibre. As described, (g) can be, together with (d), rotated (ϕ_1) around the main axis.
Further details concerning the setup are found in the description of the testing procedures to determine the cross-sectional area (Section 3.1.2) and the strain of the fibre (Section 3.1.3).
Figures 3.1, 3.2 and 3.3 display the experimental setup graphically, laying the focus on different aspects of the test setup and the digital image correlation procedure. Hereinafter, the features will be explained step by step, elucidating how they offer improvements in determining cross-sectional areas and strains during tensile tests with enhanced accuracy. The SiFiT was introduced in Figure 3.1. Features of the SiFiT setup required for strain measurements are detailed in the following in reference to Figures 3.2 and 3.3. In Figure 3.2, the overview of setup is shown for a clamping length of 20 mm. This corresponds to the largest clamping length that could be used for recycled carbon fibres within this work. The measuring length is variably adjustable, but a minimum of approximately 30 mm of spare length of the fibre is needed for handling and gripping.
It is worth mentioning that the LDS could not be operated in case of short fibres (< 20 mm). Thus, the fibres were first clamped at 20 mm to determine the cross-sectional area, the lower gripper was then opened, the upper gripper positioned correctly, and the fibre was clamped with the distance desired for the mechanical test.

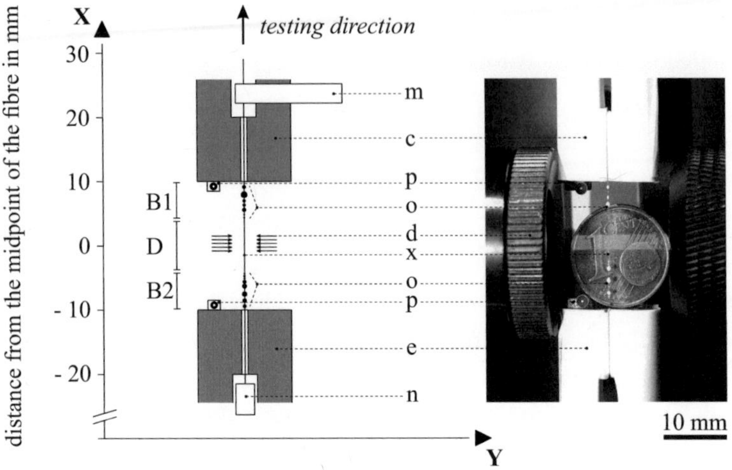

Figure 3.2: Features for the strain measurement on single fibres. B1 and B2 are regions where tracking beads are attached and the central region D is used for cross-sectional area measurements with the laser diffraction sensor (d).
Further features are: m) Sticky paper to handle the fibre; c) upper gripping finger; p) marker to track displacement of the gripper; o beads; x) carbon fibre; e) lower gripping finger; n) paper flag to align the fibre. *The coin serves as size comparison; the clamping length of the fibre is* 20 mm.

The region of interest for the single fibre tensile tests is equal to the measuring length and consist of two bead regions (B1) and (B2) and a region free of beads (D). In (D), the cross-sectional area is determined by means of the laser diffraction sensor (d) prior to testing. The sensor automatically covers a length of 2 mm along the fibre. In order to ensure error-free determination of the diameter, it is necessary to have a good, vertical alignment of the fibre with regard to the laser diffraction sensor. To handle the fibre, a paper flag (m) is attached to the upper end of the fibre. A smaller paper flag (n) serves as weight to align the fibre by gravitational pull when transferring it to the testing device. Alternatively to gravitational alignment, contact-free alignment via suction by an installed Venturi nozzle can be used. Whilst this works well for long glass fibres [109], it was disadvantageous for carbon fibres and it was decided in favour of gravitational alignment. Once the fibre is placed within the slit between the upper grippers (c), they are pneumatically closed before the lower gripper (e) is shut. After closing the grippers, it is beneficial to apply a slight pre-tension in order to prevent wrong readings of the diameter with the LDS0200 (d). The features explained so far are needed for running tensile tests focussing on the ultimate tensile strength. In contrast, for measuring strains and calculating the Young's modulus, two additional characteristics are needed. Tracking markers, here referred to as "beads", are applied onto the fibre. They consist of the adhesive *proformic C1001* by VIKO UG blended with a powder containing yellow and orange fluorescing pigments, ash and carbon powder. The amounts of these powdery constituents were optimised in order to provide a distinctive optical pattern to be used for digital image correlation in the experiment. Thanks to surface tension, the adhesive forms spherical markers. The adhesive-powder-paste cures within second when applying ultraviolet light. UV light serves two purposes in the SiFiT procedure: firstly, it cures the tracking beads; secondly, it increases the stability of the digital image correlation. At least five individual droplets (beads) should be applied in both regions, (B1) and (B2), to obtain stable measurements.

The gripping areas of the gripping fingers are rectangular. No chamfer was machined into the opposing edges. Thus, the distance between the grippers, as well as the gauge length can be reduced until the gripping fingers touch. This is defined as a clamping length of 0 mm. In practice, a minimal clamping length of 1 mm was realised for determining fibre strength. For determining the Young's modulus of single fibres, clamping lengths of less than 20 mm are not recommendable, as regions B1 and B2 would then coincide and accuracy would be low. The maximal length of the SiFiT is only affected by the specifications of the universal testing machine and the available length of the fibre and can be adapted to the user's demands.

The upper part of Figure 3.3 is a schematic illustration of the setup, the bottom part shows a corresponding image taken during an experiment. This illustration is needed to elucidate the pair-wise correlation of bead in the digital image correlation procedure, described in Section 3.1.3.

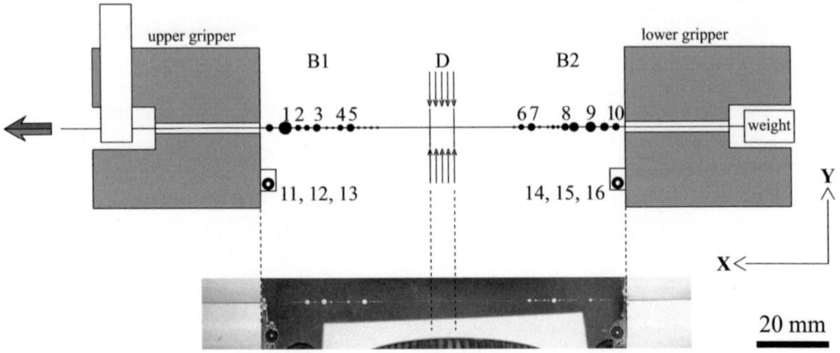

Figure 3.3: Zoom on the region of interest for digital image correlation: Beads 1 to 10 are defined for determining strains between the upper region (B1) and the lower region (B2). Beads 11 to 16 are used for indirect strain measurements. Beads 1, 10 and 11 set a triangle needed to define a coordinate system with vector $\overrightarrow{P_{10}P_1}$ being the positive x-direction. *Here, the depicted clamping length of the fibre is* 50 mm.

The described experimental setup meets the requirements listed at the beginning of this chapter. Furthermore, it takes equation 2.1 and the motivation presented in Chapter 2.1.4 into account.

3.1.2 Determination of cross-sectional areas of carbon fibres

Experiments to determine the Young's modulus by applying the SiFiT method were run controlled by nominal stress and consequently, the cross-sectional area of the fibres had to be known before the mechanical test started. This excludes the possibility of SEM and optical microscope investigations: side views could only be used to determine an apparent diameter, again assuming circular fibres. To obtain the real fibre shape with these methods, the fibre would have to be embedded and polished and could not be used for mechanical tests afterwards. Microsections were thus only used for subsequent investigations and quality assessment. Derived from the state of research, only laser diffraction sensors offer possibilities to be used for in-situ approximations of real fibre shapes [1].

In the state of research and in particular in Table 2.2, it was demonstrated that carbon fibres can significantly differ from perfectly circular shapes, more likely being oval or even kidney-shaped. Fibres of type PANEX 35 are best assigned to categories labelled as *circular* (a) or *elliptical/oval* (c). As a consequence, a procedure to approximate both shapes by means of laser diffraction measurements had to be developed. The following procedure is based on the assumption that oval shapes of carbon fibre can be simplified by elliptical approximation, as introduced in Figure 2.5.

Derivation of a procedure to approximate the cross-sectional area

Generally speaking, laser diffraction sensors or laser scanning devices will yield a distance between two edges of an object. If the object is not point-symmetric, the measured distance d', the so-called apparent diameter, will vary as a function of the angular position. This is schematically shown as top view of a fibre, with the system being rotated by 20° in respect to the normal plane in Figure 3.16.

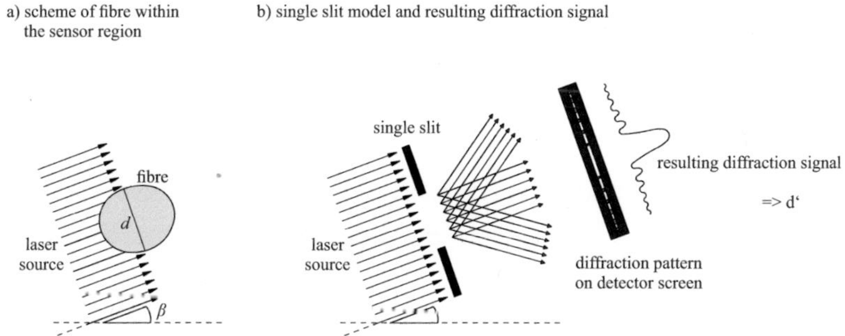

Figure 3.4: Principle of individual diameter measurements to approach an oval shape. a) simplified presentation of the laser waves interacting with a fibre rotated by an angle $\beta = 20°$ in relation to the reference position of the laser. b) Scheme of the laser diffraction principle: the fibre is approximated by a single slit, resulting in a diffraction pattern on the detector screen. The diffraction pattern is then transferred to a signal from which the apparent diameter d' can be calculated.

The signal d' can be transferred to a distance d. It is then assumed that a relative rotation between the object (fibre) and sensor will give different readings as a function of the angular rotation.

Reflecting basic geometrical consideration, plotting the readings of the diameter over the angle of rotation will yield a straight line for a perfect circle and a regular sine-shaped curve for an ellipse, as demonstrated in Figure 3.6 for two different ellipses and a circle.

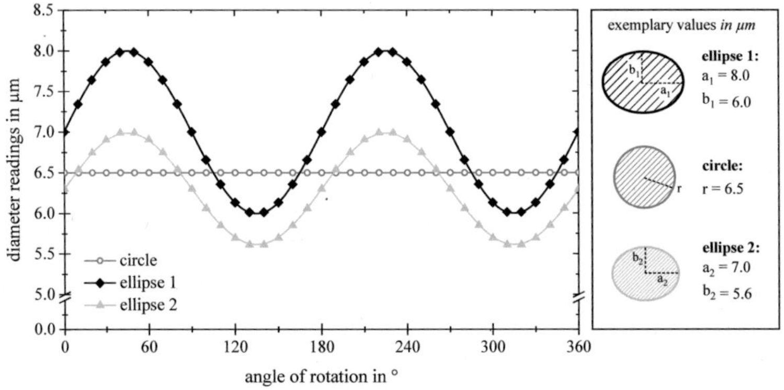

Figure 3.5: Theoretical examples for the approximation of a circle and two ellipses from diameter readings as a function of the angular position of a laser diffraction sensor.

The greater the difference between the semi-major (a) and the semi-minor axis (b) of the ellipse, that is, the higher the distortion of the elliptical shape, the higher the amplitude of the sine function. Additionally, it can be deduced, that when using laser-based techniques, a rotation of 180° suffices, as measured values of the apparent diameter will repeat afterwards [1].

It can be inferred that the shapes of elliptical carbon fibres can be approximated by taking several measurements as a function of the angle of rotation and using a sine-fit for receiving the two semi-axes. The following procedure was developed to gather the experimental data and to calculate a non-linear fit of least squares that yields semi-major and semi-minor axes.

Application of the approximation procedure

As described in Section 3.1.1, single fibres were clamped in the SiFiT (Figure 3.1). A pretension of 5 mN was applied onto the fibre to ensure straight fibre alignment and correct readings. Individual measurements with the laser diffraction sensor were taken in angular steps of 10°. Each measured value obtained in this way was plotted as a function of the angular position, shown in Figure 3.6. It should be noted that neither the epicentres of all individual sections, nor the centre of the fibre and the central axis of the test rig necessarily coincided.

Due to geometrical constraints of the setup, the angle could only be varied up to 140°, as the sensor then collided with the frame of the testing machine. As a consequence, an angle of 40° remained as a blind spot. Despite this constraint of the setup and the missing data, a procedure was found to compute the cross-sectional area with high accuracy.

In total, 15 individual diameter measurements (d_i') were taken in the middle of the clamping length. The diffraction sensor automatically scans and averages the values within a length of 2 mm along the fibre. The manufacturer guarantees a correct functionality down to diameters of 5 µm with an accuracy of ± 0.01 µm but also smaller diameters should allow for an error-free measurement. Exemplary, the measurements of d' in respect to the angular position are plotted in Figure 3.6.

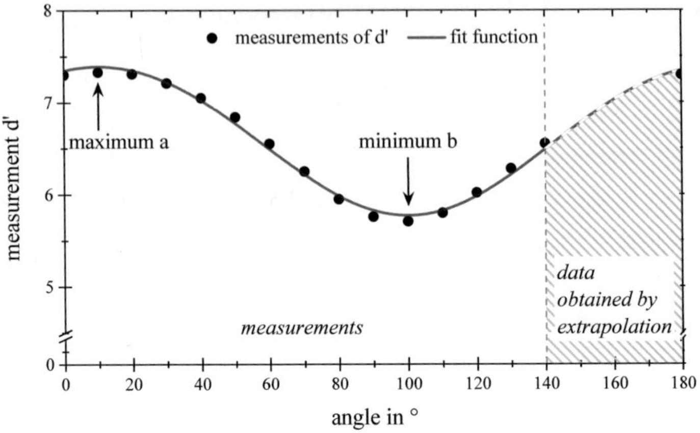

Figure 3.6: The described fit function applied to individual measurements of apparent diameters d' of a real fibre as a function of the rotational angle.

The figure not only displays the measured values but also the fitting curve obtained by a fit of least squares based on Equation 3.4 of the measured data. The following derivation was used to determine the cross-sectional area based on an elliptical assumption. Figure 3.7 gives a graphical description of the geometry.

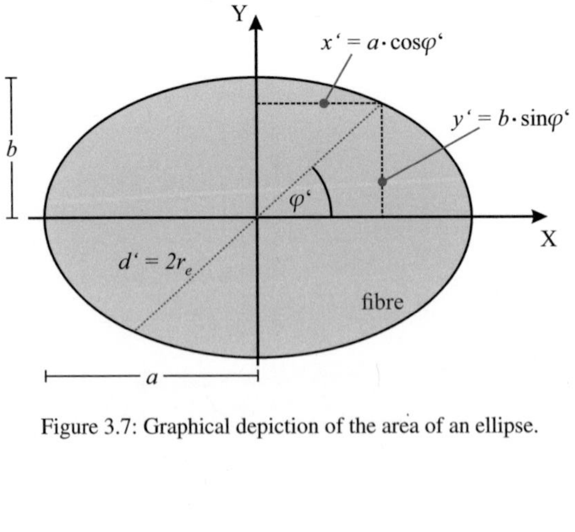

Figure 3.7: Graphical depiction of the area of an ellipse.

The calculation makes use of the equations:

$$x' = a \cdot cos\varphi' \tag{3.1}$$

$$y' = b \cdot sin\varphi' \tag{3.2}$$

and of Pythagoras' theorem:

$$(2 \cdot r)^2 = d'^2 = (2 \cdot x')^2 + (2 \cdot y')^2 \tag{3.3}$$

The semi-axes a and b can then be determined by:

$$d' = \sqrt{(2 \cdot a \cdot cos\varphi')^2 + (2 \cdot b \cdot sin\varphi')^2} \tag{3.4}$$

Subsequently, the non-linear fit of least squares is run and yields semi-major and semi-minor axes of the approximated ellipse. These axes are then used to calculate the area of the elliptical cross-section (A_{fibre}) according to:

$$A_{fibre} = \pi \cdot a \cdot b \tag{3.5}$$

As Figure 3.6 shows, the fit can cope with the lack of data in the region from 140 to 180°. Maximum a and minimum b of the fit do not necessarily coincide with measured values. The procedure is likewise applicable for oval and circular fibres, as for the latter, both axes will attain the same value. However, due to the physical principle of laser diffraction, this method can only be applied to opaque fibres like carbon fibres and natural fibres. Transparent or translucent fibres such as glass and basalt cannot be investigated with a laser diffraction sensor.

The geometrical constraints of the experimental setup did not allow to find the thinnest section along the fibre prior to testing. Thus, the lengthwise middle-point of the fibre was chosen to determine the cross-sectional area.

3.1.3 Direct strain measurements in single fibre testing

Digital image correlation

Contacting methods of strain measurement [118] are not applicable on single carbon fibres. Instead, non-contacting methods like digital image correlation are the method of choice for investigating fibrous objects. Still, it is more complicated to apply DIC on single fibre level than on bulk, macro-scale specimens. Onto the latter, a unique stochastic pattern can be applied by spray-coating the specimen with black and white paint, shown in Figure 3.8 a). As thin fibres themselves do not exhibit a optical distinct pattern, (Figure 3.8 b)), tracking markers, hereafter referred to as beads, need to be applied (Figure 3.8 c)). In addition, the reader may refer to Figure 3.10 where a SEM image of these beads is shown.

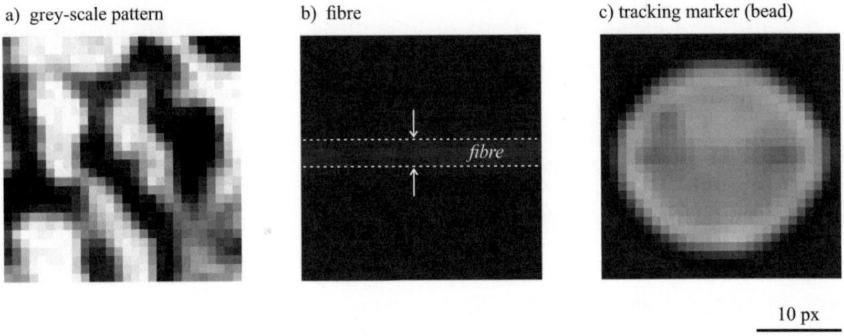

Figure 3.8: Digital image correlation in tensile testing. a) To use DIC in macro-scale tensile testing, a stochastic grey-scale pattern is sprayed on the specimen; b) fibres do not show any distinct features to apply DIC; c) by attaching tracking markers (beads) to the fibre, DIC can be used on single fibre level. All squares represent a 28 × 28 pixel section, corresponding to approximately 0.25 mm.

For explaining the measuring procedure in more detail, a zoom on Figure 3.2 is helpful. Whilst most features have been explained before (Section 3.1.1), additional information on the strain measurement by digital image correlation is needed. The exact amount of tracking beads applied on the fibre could not be defined purposely. Instead, a small amount of the polymer-pigment-mix is put onto the fibre using a thin polymer-coated wire. It is spread over a distance of approximately 5 to 8 mm. By surface tension, beads evolve, but the exact position and size cannot be manipulated by the user. The result can be seen in Figures 3.9.

A zoom on individual beads and facets (Figure 3.9) highlights typical characteristics and sizes of the facets.

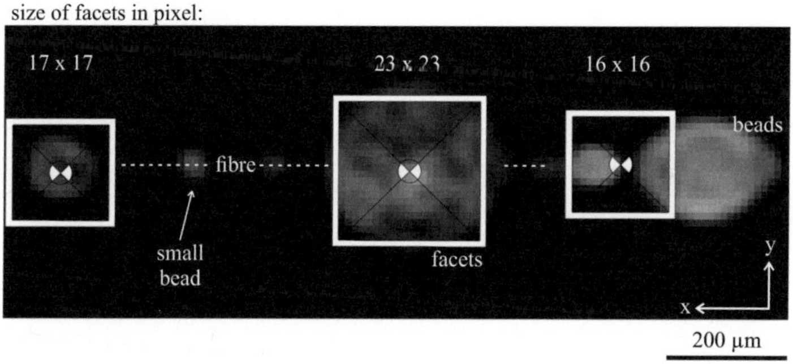

Figure 3.9: Zoom on exemplary facets. The individual pixels of the image become visible. It is evident, that the fibre itself does not show a distinctive pattern while the beads possess a unique grey-scale information. *Published by the author in* [110].

Referencing Figures 3.3 and 3.9, the data collection is explained: to determine strains, the following post-processing was executed in the DIC-software GOM *Correlate Professional 2017*. A series of images was imported, and on the first frame, corresponding to a reference level, initial facets were defined. Facets were applied manually to the most distinct tracking beads. This process was aided by the software, providing the so-called "pattern quality" for each square facet. Facets were placed on beads where the quality index was maximised. In total, 16 facets were used for each evaluation, as shown schematically in Figure 3.3.

Facets 1 to 5 represent the upper region of the fibre ($B1$), 6 to 10 the lower region ($B2$). The facets labelled 11 to 13 and 14 to 16 are connected to the gripping system. They are needed for aligning the strain determination with the testing direction and to demonstrate the difference between direct strain measurement (DSM, on the fibre) and indirect measurements (ISM, on the gripping system). Facets 1, 10 and 11 are needed to create a virtual coordinate system in the x-y-plane, facing in testing direction, Figure 3.3. The more accurate facets 1 and 10 are centred in respect to the fibre axis, the better the alignment of the strain measurement with the testing direction. Centring of facets 2 to 9 is less important, as only strains in X-directions were later considered. For facets 2 to 9, the main target was to achieve a high pattern quality index. It is worth mentioning, that the facets do not necessarily coincide with the centre of the bead, they can be shifted along or transverse to the fibre axis. Under some conditions, it can be favourably to place the centre of the facet between two beads (Figure 3.9).

After the manual definition of sixteen facets in the first frame, a GOM script evaluated positions of all facets in the subsequent images. Using DIC, strains in X-direction were calculated among nine individual pairs of beads, defined as follows. Firstly, the pairs were taken from the outermost to the innermost beads (1/10; 2/9; etc.). Secondly, beads were compared from top to bottom (1/6; 2/7; etc.). Thereby, several benefits could be achieved: faulty or lost facets could be detected and replaced. Tracking nine separate distance measurements in the manner described, heterogeneously distributed strains would be unveiled. Furthermore, this pairwise evaluation was needed to ensure that the polymer tracking beads did not influence the stiffness of the fibre. Thirdly, facets were applied on the grippers itself. A pairwise observation of the facets (11/16), (12/15) and (13/14) was used to track the movement of the grippers. Measurements on the grippers are useful to discuss effects of the systems compliance.

In total, each experimental run provided twelve strain readings per load step and image; nine for DSM and three for ISM to study the differences. The total account of strain measurements varied as a function of tensile strength, as it depends on the amount of images gathered before fibre failure. Starting with the initial facets in the reference image, displacement and deformation of the

16 facets were tracked automatically for a series of images, obtained at different load levels.
With the given setup and camera, a length of 10 mm corresponds to approximately 1050 pixel, or 105 pixel mm^{-1}. As additional information, Figure 3.10 shows four exemplary beads as observed in a SEM image and how they might appear in the digital image correlation images. The photograph is not to scale.

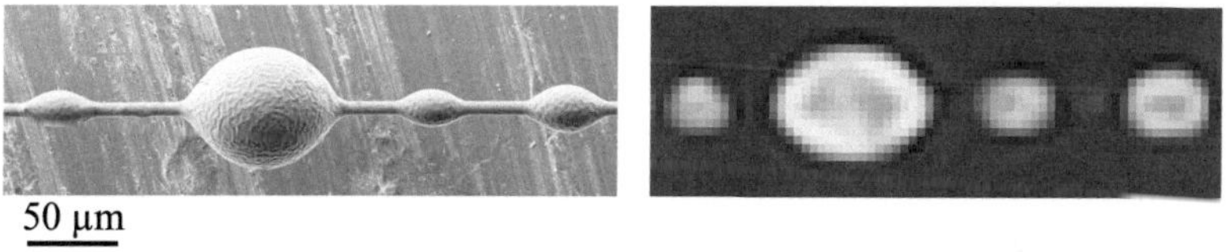

Figure 3.10: A scanning electron micrograph of tracking beads (left, *SEM/IAM*) and a similar image detail as it appears in digital image correlation (right) during the tensile testing procedure *The right figure is not to scale.*

Commonly, strain to failure, strength and the Young's modulus are measured in one and the same experiment. Applying DIC on fibre level, this becomes inopportune. Instead, the experiments were divided: tests to obtain the strength of single fibres were executed continuously until failure; Young's modulus of single fibres was determined discontinuously in a step-wise procedure; strain to failure was not under investigation. Thus, experimental details for both realised experiments are outlined separately in the following.

3.1.4 Tensile testing procedures

Determination of tensile strength

For experiments dealing with the ultimate tensile strength, the strain of the fibre was not considered given the inaccuracies in the indirect determination of strain discussed before [111, 113, 115, 116]. No beads were applied in test concerned with the tensile strength.
Standards for single fibre testing remain vague concerning a definite strain rate or a testing speed [111]. It is instead stated that failure should occur within

30 s [113]. Due to the failure behaviour best described by a Weibull distribution, tensile strength and thus time to failure at constant speed vary significantly. Applying the recommendation of the standard would thus inevitably result in a large variation of testing time. The benchmark has therefore been customised and the target was to reach a tension of 4 GPa within the recommended 30 s, whether or not failure occurred within this time. The value has been chosen as it was the approximate average of the tensile strength. In [314], it was shown that HT fibres are not sensitive to the strain rate.

Determination of the Young's modulus

One of the major goals of the work at hand was to develop a procedure to determine the strain on single fibre level. In contrast to common procedures, strains were to be measured directly on the fibre. This requires some amendments to standard procedures. Tracking markers, here called beads, had to be applied on the fibre. These served as markers in the later evaluation by means of digital image correlation. To fully benefit from digital image correlation properly, an alternative testing procedure was designed: instead of loading the fibre continuously until failure, a step-wise procedure was introduced and the main part is graphically shown in Figure 3.11. Thereby, it was possible to take high-resolution images on each step. As the exposure time of the digital camera had to be set to 8 s, continuous loading was out of the question.

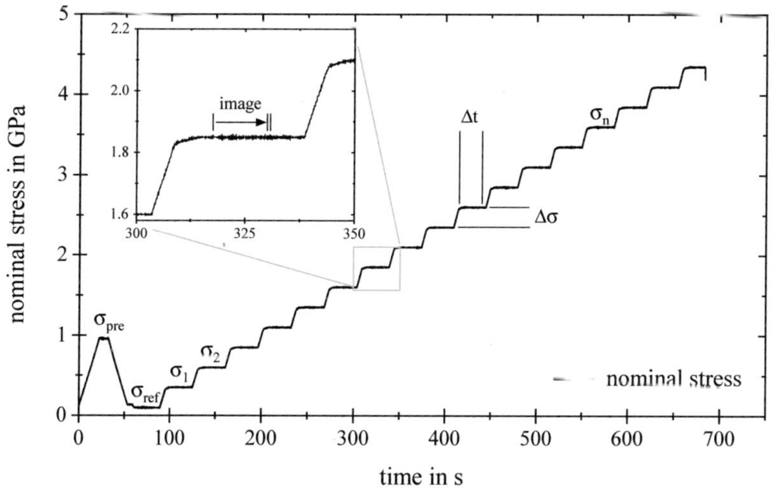

Figure 3.11: The load cycle as used for the determination of Young's modulus according to the SiFiT procedure. A step-wise loading was needed to use high-resolution digital image correlation.

Strictly speaking, each test cycle consisted of two parts, with Figure 3.11 displaying only the second. Prior to that, the cross-sectional area was determined in a first run as described before. This pre-test was then aborted to allow the input of the cross-sectional area in the *testXpert* software. Thus, the main experiment could be controlled by nominal stress instead of load. The cycle started with a relatively high pre-cycle stress σ_{pre}. If fibre slippage was to occur, it could be detected in the pre-cycle and the experiment would be aborted. After the pre-cycle, the main experiment started on a reference nominal stress level σ_{ref} and the stress is then increased in steps of $\Delta\sigma$. A dwell time Δt was to applied to allow for a stabilisation of the force signal before manually triggering the camera. Settings that proved to be useful are listed in table 3.1.

Table 3.1: Time and nominal stresses for the experimental procedure

dwell time Δt	nom. pre-cycle stress σ_{pre} in MPa	nom. reference stress σ_{ref} in MPa	nom. stress increment $\Delta\sigma$ in MPa
30	1000 – 1500	50 – 100	150 – 500

The image taken at σ_{ref} was used as reference image for digital image correlation, run according to the afore described procedure. All stresses were nominal values referring to the initial cross-sectional area determined by laser diffraction. The amount of steps reached and images taken was depended on the tensile strength of the individual fibres. Tensile strength was not considered in the step-wise experiments.

Determining the Poisson's ratio

Albeit the Poisson's ratio is an important elastic property of materials, only few publications on possible procedures to determine it for technical fibes are available [171, 233]. As the SiFiT setup and procedure presented here allow for an enhanced determination of strain and cross-sectional area of single fibres, it fulfils all requirements to determine the Poisson's ratio. The fact that the experiment needs to be run stepwise to produce high-resolution images is even more advantageous when determining the Poisson's ratio. In general, minor variations of the fibre's cross-sectional area can occur along the fibre (see also Figure 3.15). While they are negligible in terms of the standard tensile tests, they need to be considered when determining the Poisson's ratio. Here, it has to be ensured that variations in the apparent diameter are caused by a lateral contraction rather than a shifted focus of the laser. In the SiFiT setup, only the upper gripper, attached to the crosshead, is moved. To counteract this movement, a manual tracking of the focus of the LDS0200 was executed: a depth calliper was attached to the standard setup, and the LDS was shifted by half the displacement of the crosshead on each step. Due to this minor change in the experimental procedure, the Poisson's ratio was determined in an independent experiment. The lateral contraction per step was rather low. Thus, the load increments per step ($\Delta\sigma$) were increased. A procedure alike to that for determining the Young's modulus (cf. Figure 3.11) was executed. Here, each dwell time was held until it was manually confirmed that all substeps bad been executed successfully. The substeps were: taking an image for strain correlation, shifting the beam focus of the LDS and acquire readings of the apparent diameter. As for most of the other tensile experiments, 15 individual readings of the apparent diameter were taken. Again, the elliptic fit was used to determine the cross-sectional area. However, as a lateral contraction had to be considered, the cross-sectional area was then translated to the apparent diameter of a regular circle, here labelled $d_{app,0}$. For calculating Poission's ratio, the following equation is applied:

$$\nu = -\frac{\varepsilon_d}{\varepsilon_l} = -\frac{\Delta d}{d_{app,0}} \cdot \frac{l_0}{\Delta l} \tag{3.6}$$

Here, ε_d is the strain perpendicular to the fibre direction, ε_l the strain in the fibre direction, Δd variation of diameter, l_0 the initial length and Δl the variation in length.

3.2 Method validation and reliability studies

The described procedures to measure the cross-sectional area and the strain of carbon fibres are novelties. Before comparing them to the related state of research in single fibre testing, several experimental studies are presented and discussed here to evaluate applicability and reliability of the novel procedures. They are mostly independent of the specific carbon fibres but aim on general verifications.

3.2.1 Reliability of strain measurements

The reliability of strain measurements was investigated in terms of the quality of the digital image correlation and of possible effects due to the application of tracking beads.

A) Artificial image distortion – accuracy of DIC

It is legitimate to ask whether the strain determination via digital image correlation is a reliable procedure. For conformation, the following virtual experiment was run.

Experiment
From the initial images of all real experiment, six were randomly picked. These images were graphically distorted by the factors listed in Table 3.2.

Table 3.2: Factors used for virtual image distortion

distortion factors in %													
0.1	0.2	0.5	1.0	1.5	2.0	3.0	4.0	5.0	6.0	7.0	8.0	9.0	10.0

A routine according to the real experiments was executed: ten facets were positioned in the DIC software on beads present in the initial frame. Digital image correlation was then run on series of distorted images. An exemplary detail, here for the largest distortion, is shown in Figure 3.12.

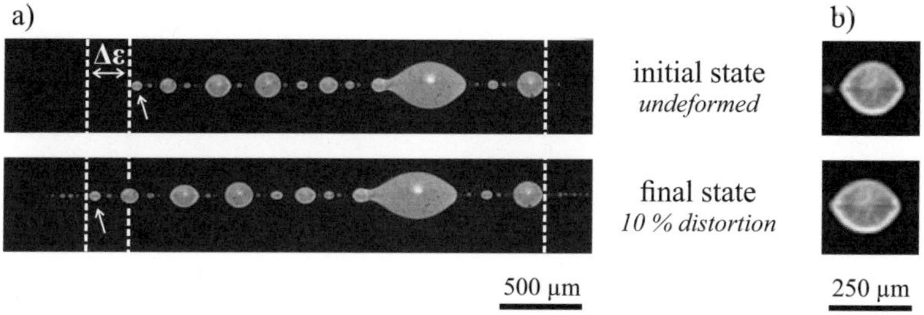

Figure 3.12: Distorted facets to confirm the accuracy of the DIC routine. The distortion of a series of beads is shown in a); the deformation of a single bead is presented in b).

In a), the white line on the right indicates the base line. The upper series of beads presents the initial, undeformed state. The bottom "fibre"[1] is in its final state, deformed by 10 %. Here, the beads and their positions are clearly distorted, as indicated by the white arrow. In b), a higher magnification of a single bead in both states is shown. The distortion of the bead is clearly visible for the naked eye, but as Figure 3.13 shows, even high distortions are tracked by DIC with highest accuracy.

Results

Figure 3.13 is a plot of the averaged values given as output by DIC computation versus its designated distortion. The fit line has a Pearson-coefficient of almost 1. Standard deviations were too low to be displayed as error bars. Instead, they are graphed with respect to the right Y-axes.

[1] The inverted commas are used her to indicate, that "strain" and "fibre" are equivalents of the real experiment.

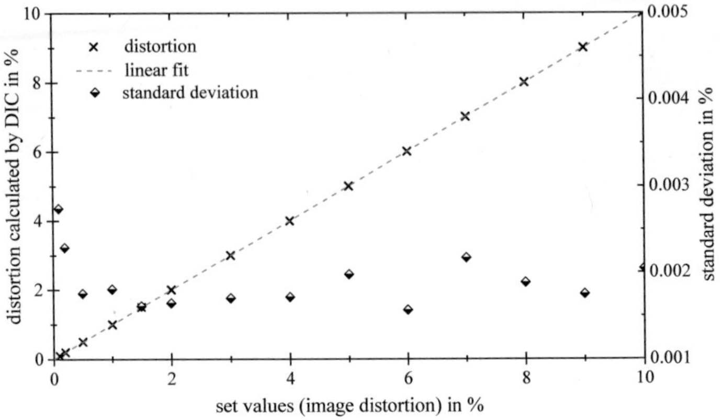

Figure 3.13: Verification of the digital image correlation procedure: initial images were distorted using computer software and "strains", subsequently calculating strains by means of DIC.

Conclusion concerning the accuracy of DIC

The virtual experiment verifies the high accuracy of the digital image correlation routine followed in the experimental part. Even beads distorted by factors up to 10 % are recognised error-free and the "strains" between them are calculated correctly. As carbon fibres can typically not be strained more than 2 % before they fail, the DIC procedure can be expected to work reliably.

B) Effects of tracking markers

As explained, cameras cannot be used to carry out digital image correlation directly on thin single fibres. For the strain experiments using DIC, polymer beads were attached to the fibre as tracking markers. One might argue that this corresponds to the creation of a single fibre composite. Inevitably, a question arises: do the tracking beads affect the results of the strain measurement by stiffening the fibre? This paragraph will confirm that this is not the case on base of an evaluation of the experimental data.

Experiment

The experiment relies on real data obtained in tensile test with carbon fibres of 35 mm. However, type and material of the fibre are of no importance in this evaluation. The approach should be considered general and independent of the determination of the actual fibre properties, Section 5.2. A comprehensive evaluation on glass fibres is also found in [109].

If the beads had a stiffening effect, then the innermost pair would always yield the lowest, the outermost pair the highest results for the Young's modulus. Vice versa, strain measurements would show the opposite trend.

Results and discussion

By default, beads were paired in a way that pair 1/10 represents the pair with the greatest distance, 5/6 a pair which lies more to the centre of the fibre. This is exemplary shown for three fibres in Figure 3.14.

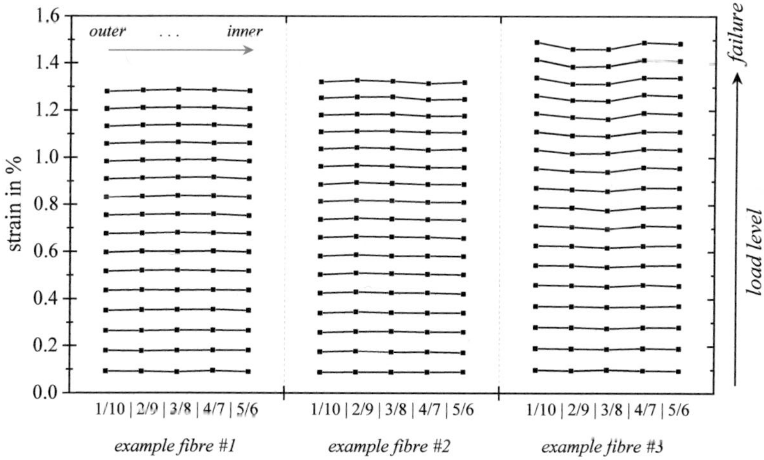

Figure 3.14: Strain measurements as a function of bead coupling for three exemplary fibres. If the beads had a stiffening effect, a trend towards larger strains for the innermost pair (5/6), would be observed.

Strains as determined by digital image correlation are plotted versus the pair of beads. In each column, the left data point is the strain measured with pair 1/10 (outermost pair). From bottom to top, the strain levels can be counted. The amount of lines/levels corresponds to the load levels reached before failure occurred and can thus vary. It is evident, that there is no trend which would indicate a stiffening effect. If the fibre was stiffened/reinforced, the inner pair (5/6) would yield the highest strain readings, which it does not. Albeit scatter is found between different bead pairs, no trend occurs. Fluctuations become more distinct for higher load levels, as seen in *example fibre #3*. However, the fluctuations are assigned to experimental limitations in the accuracy of the digital image correlation and are not a systematic error caused by possible stiffening. For the sake of clarity, only three examples are displayed graphically. However, a larger data set was examined: twelve fibres were investigated in the same way and none of the results indicated any dependency between strain reading and bead position. In conclusion, no stiffening effect due to the application of polymer tracking markers on the fibres was found.

3.2.2 Reliability of determining the cross-sectional areas

In this section, three aspects are discussed to determine the reliability of the elliptical approximation function. Variation of the cross-sectional area along the fibre are investigated, the elliptical assumption is compared to circular analogues and the calibration of the laser diffraction sensor is under investigation.

Cross-sectional areas along the fibre

Carbon fibres can vary in shape and size along their axis. With the SiFiT setup described earlier it was not possible to identify the minimal cross-sectional area of the fibres. Instead, an arbitrary position was chosen. It is thus mandatory to evaluate to what extent variations of the cross-sectional area along the fibre occur.

In different experiments, the cross-sectional area was determined according to the SiFiT procedure. Clamping lengths and distances between two measurements were varied, with the former being maximal 50 mm and the latter being minimal 1 mm. For demonstration, several fibres have been investigated lengthwise, taking measurements at different positions along the fibre. Results are presented in Figure 3.15.

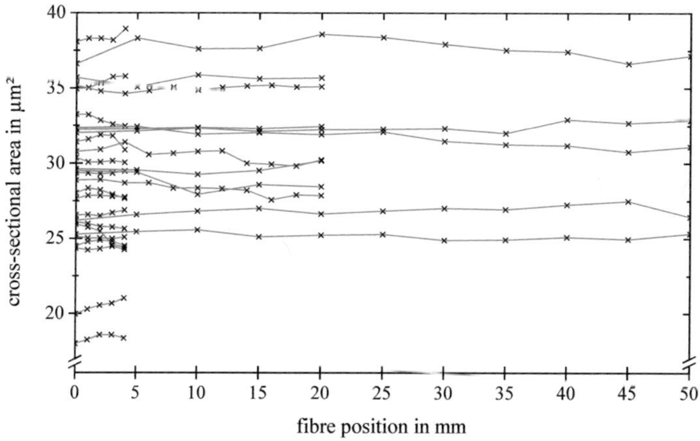

Figure 3.15: Determination of the cross-sectional area at different position on a clamped fibre. Each line represents a single fibre; each × a measured result.

Comparable to [191], each line corresponds to one individual fibre; each cross to a measured result in relation to the measuring position. Distinct differences of the cross-sectional area are found between individual fibres. Variations are also found along a single fibre. However, these remain low enough to postulate that a single measurement in the centre of the fibre's clamping length prior to mechanical testing is sufficient.

Computer-aided approximation: elliptical and circular assumptions

The novel approximation of cross-sectional areas based on elliptical fitting of laser diffraction data has not been reported in literature by others than the author yet. It is often assumed that carbon fibres show perfectly circular cross-sections. The novel procedure was intended to be more accurate than common circular approximations. However, the disadvantage of laser measurement (Section 3.1.2) is that they cannot be verified in-situ as the real fibre shape remains unknown. To prove the reliability of this novel strategy, a computed-aided procedure analogous to the laser technique was used and is presented in the following (cf. Figure 3.16).

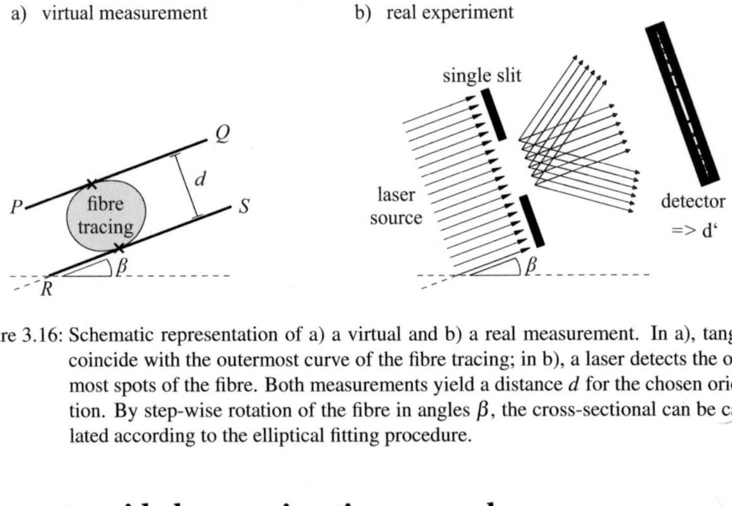

Figure 3.16: Schematic representation of a) a virtual and b) a real measurement. In a), tangents coincide with the outermost curve of the fibre tracing; in b), a laser detects the outermost spots of the fibre. Both measurements yield a distance d for the chosen orientation. By step-wise rotation of the fibre in angles β, the cross-sectional can be calculated according to the elliptical fitting procedure.

Computer-aided approximation: procedure

Using a micrograph of embedded fibres, ten arbitrary fibre shapes were manually traced in a image-processing software and their cross-sectional areas (Figure 3.17, A to J) were computed to obtain the nominal value of each geometric body. For comparison and in accordance with the real experiment with the laser diffraction sensor, a set of two parallel tangents ($\overline{PQ}$ and $\overline{RS}$) was drawn in a way that they coincided with the outermost curve of the fibre tracing, illustrated in Figure 3.16. The set of parallel tangents was then rotated by increments of 10°. At each angular step, the absolute distance between the two tangents was determined.

In total, fifteen shapes, shown in Figure 3.17, were evaluated. The objects A to J correspond to tracing of real fibres found in microscopic observation. Objects i to v are geometrical simplified representations of fibre shapes to be used as comparisons. The latter group is used to explain the potential and limitations of the method. These artificial geometries consist of: a kidney-shaped (i, for instance [140, 141, 154]) and a cross-shaped (ii, [153]) geometry, representing complex fibre shapes found in literature, presented in Table 2.2. The octagon (iii), ellipse (iv) and circle (v) can be considered approximations of more regular carbon fibres. It is very unlikely that real fibres show the high degree of symmetry presented in shapes i to v; they are rather used to verify the procedure. Hollow fibres [12, 15] and more complex shapes [14, 16], introduced in Table 2.2, are not investigated here.

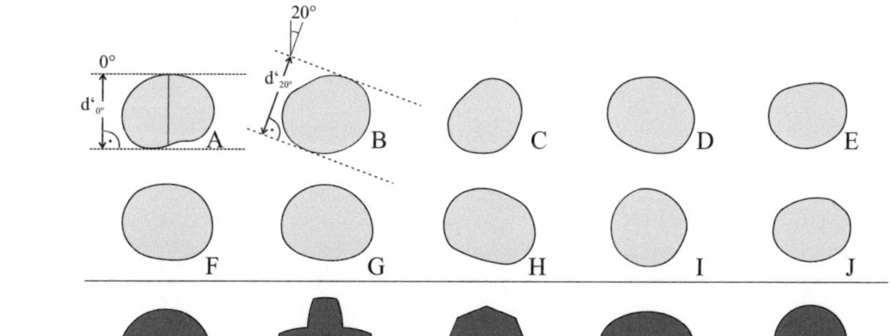

Figure 3.17: Tracings of ten real fibre shapes (light grey) and five simplified shapes: kidney-shaped, cross-shaped, octagonal, regular elliptical and circular. All fifteen fibres were measured according to the scheme in Figure 3.18. Fibre A and B are used to display the approximation procedure.

As the fibres were embedded arbitrarily, the 0°-direction equals to a random selection of the coordinate system.

The flow chart in Figure 3.18 demonstrates how the fifteen geometric tracings were used for verifying the experiment. Per geometric body, the mimicking of the real experiment yields 18 independent readings of the apparent diameter as a function of the angular position. The 0°-direction was chosen arbitrarily for the real fibre tracings and is aligned with the axis of symmetry for all regular bodies. The set of 18 measurements per fibre is considered the data base.

The Matlab-script was run for a series of either 15 or 18 individual measurements, respectively. For comparison, circular areas were calculated. There, cross-sectional areas were calculated by picking values from the database as diameter. Either individual values were taken or, alternatively, 3 or 15 individual measurements were averaged. Firstly, single measurements were taken by considering either the maximum, the minimum or a random value of the database. The random value corresponded to procedures where only one measurement per fibre is taken arbitrary. For instance, this is the case in the testing protocol of *Dia-Stron's* testing device LEX820/LDS0200. Secondly, an average of three values rotated by 60° was chosen as this is an approach standing to reason. Thirdly, a set of 15 measurement, rotated by 10° was used to obtain a comparison to the newly developed elliptical approach. The result "true value" was computed by image processing and was defined as the nominal value. All approximated results were compared to this nominal value.

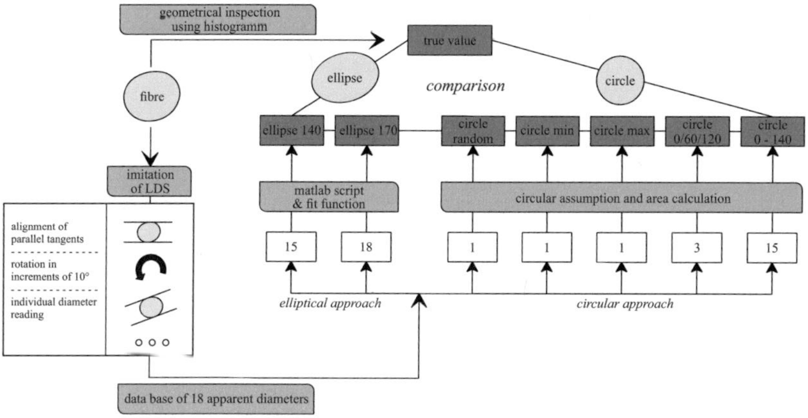

Figure 3.18: Flow chart of a CAD-based evaluation of the approximation of a elliptic fibre shapes.

Summarising, the following values were calculated:

- **Ellipse 140** was calculated by the elliptical fit presented in Section 3.1.2. Apparent diameters were taken in steps of 10° from 0 to 140°. This corresponds to the constraints caused by the experimental setup of the SiFiT.
- **Ellipse 170** is equivalent to the procedure for *ellipse 140* but considers angles up to 170°. This corresponds to a half circle. Due to the working principle of laser diffraction and the gathering of apparent diameters, it equals to covering a complete circle.
- For **circle random**, one apparent diameter for each fibre tracing was randomly picked and its circular cross-sectional area was calculated.
- The maximum (**circle max**) and minimum (**circle min**) apparent diameter of each fibre tracing were used for calculating a circular cross-sectional area.

- For **circle 0/60/120**, the arbitrary 0°- direction was picked as the starting point. The apparent diameters at 0, 60 and 120° were averaged to calculate the cross-sectional area of a circle.

- **For circle 0-140**, the same values as for *ellipse 140* were picked. Instead of using the elliptical fit, the 15 apparent diameters were averaged and a circular area was calculated.

Computer-aided approximation: results

The results of the variations when evaluating different data sets are displayed in Figures 3.19 and 3.20, the latter zooming on the most relevant data selections. Labels *A* to *J* refer to tracings of real fibres; *i* to *v* to the geometrical references.

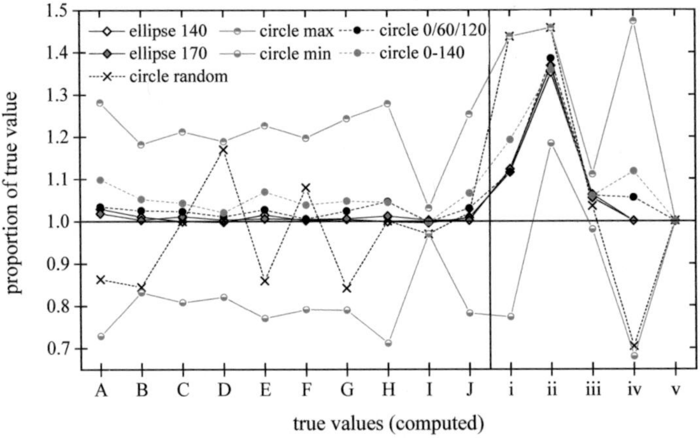

Figure 3.19: Cross-sectional areas calculated by different elliptical and circular approaches in comparison to the true value computed directly via image processing. *The lines are not supposed to show a trend but to improve visibility.*

Figure 3.19 reveals that some of the circular assumptions lead to poor results compared to the true value. The assumptions *min* and *max* are worst case scenarios and can be seen as threshold. Assumption *random* lies within these borders but shows large deviations. Although it is possible that a suitable

measured value is taken at coincidence, large errors must be expected here. These three circular calculations show very large deviations for the actual carbon fibre shapes and on this base, none of them can be recommended. Nonetheless one should bear in mind that testing devices such as *Dia-Stron's* LEX820/LDS0200 correspond to the procedure equivalent to "*random*".
After exclusion of the calculations *max, min* and *random*, a closer look on the remaining procedures is shown Figure 3.20.

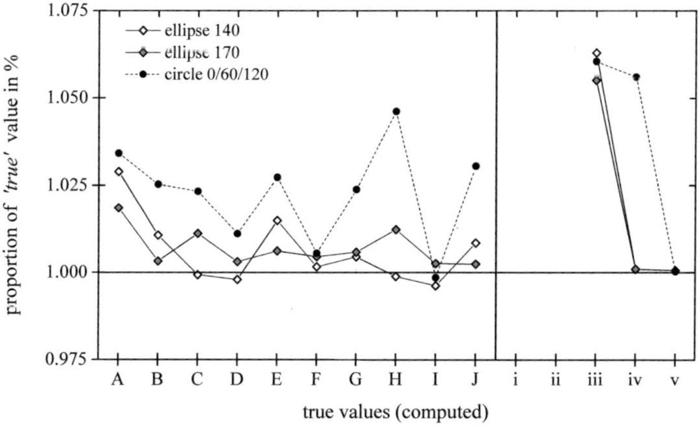

Figure 3.20: Zoom on the most relevant procedures shown in Figure 3.19. Values for i and ii are not displayed for the sake of clarity.

Comparing different approximations with the true value of the cross-sectional area, calculated by image processing, Figure 3.19 and 3.20 demonstrate the following: both elliptical assumptions show deviations which are significantly lower than almost any of the circular assumptions. Only the random assumption might be close to the true value, but as Figure 3.19 shows, the hits for fibre shapes H and J were just by chance and are not reproducible.
The following conclusions can be drawn from the virtual experiment:
Laser diffraction techniques, here mimicked by a computer-aided graphical approach, tend to overestimate the cross-sectional area. Crenellations on the

fibre surface or kidney-shaped undercuts cannot be imaged by laser diffraction, leading to a "smoothed" result, as also depicted in Figure 3.32. It is thus more likely, that an individual apparent diameter is an overestimation. As a consequence, it is recommended [1] to repeat the measurement in different angular positions to reduce the variance.

Calibration of diffraction sensor

In the state of research, a detailed overview of applications of laser diffraction sensors for determining the cross-sectional areas of carbon fibres prior to mechanical testing was presented [179–183]. It was demonstrated that laser diffraction is a valuable and trustworthy tool to determine apparent diameters of carbon fibres [1, 163, 173].
Notwithstanding these facts, some studies [172, 176, 177] raised doubts concerning the accuracy of the laser-based technique in determining the cross-sectional areas of technical fibres. The work at hand underlines possible disruptive effects when using laser diffraction for the determination of carbon fibre diameters. Experiments designed to confirm the method revealed sources of error and are discussed in detail here. Similar findings were reported, stating that accuracy is high for gold-coated glass fibres [173] but decreased for carbon fibres [172]. Despite all arguments in favour of laser diffractionas as a preferable technique to measure cross-sectional areas of opaque fibres [1, 163], possible errors need to be addressed. Three independent experiments were set up and all indicate, that the laser diffraction sensor used in this work was not calibrated properly for carbon fibres.
For all experiments dealing with the calibration of the diffraction sensor, carbon fibres of type PANEX 35 were used. Images of this type of fibre are shown as photograph and SEM[2] images in Figure 3.21.

[2] Refer to Table 5.1 for information concerning the scanning electron microscopes used within the work at hand

Figure 3.21: PANEX 35 as photograph (left) and scanning electron micrographs.// SEM/IAM and SEM/LEM

Experiment 1

In cooperation with the Institute of Vehicle System Technology (FAST) of KIT, a ground and polished micro-section of an embedded bundle of PANEX 35 fibres was chosen as a random sample. The tracings of fibres in a micrograph are shown in Figure 3.22. An evaluation algorithm developed at FAST searches the image for elliptical objects, runs an elliptical fit and calculates the cross-sectional areas. Furthermore it yields a quality index. All results of the population were sorted according to the quality index. The best 54 %, a total of 76 fibres, were considered for determining the average cross-sectional area. The fibres considered for determining the average are marked with a grey dot.

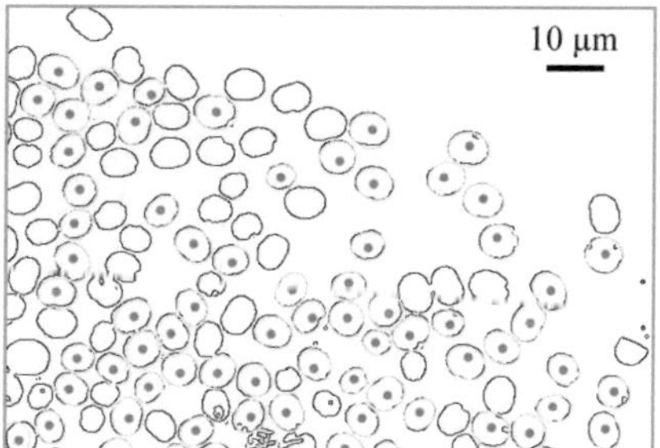

Figure 3.22: Cross-sectional areas of a polished micrograph, showing the tracing of a carbon fibre bundle.

This evaluation procedure yields a mean value of $41.25 \pm 5.16\,\mu m^2$. Even though the sample size was small, the result is understood as a first indication that the results obtained by the LDS0200 are not reliable. Results obtained by microscopic investigations deviate by approximately 25 % compared to results originally gained by means of a laser diffraction sensor calibrated with thin steel wires.

Experiment 2

The second experiment addresses the interaction of the surface with the laser beam by comparing virgin, untreated carbon fibres before and after a layer of gold was sputtered on the fibres. Step by step, gold was sputtered onto virgin carbon fibres. Each step comprised three runs, with the fibre being rotated by an angle of 60° after each run to guarantee a thorough coating. As demonstrated in Figure 3.23, the first layer of gold raised the cross-sectional area by approximately 17 %. There is no evidence to suggest that this major increase in cross-sectional area could be attributed to an increased fibre "diameter". It would contradict any experience to assume that a gold layer of a thickness of 0.5 µm could be deposited by the applied sputter parameters. The layer is more likely around 5 nm in thickness. Furthermore, a second and a third round of sputtering on the already sputtered fibre caused only a slight increase of the fibre diameter. This might be explained by an increase in thickness, but could also be in the range of variation.

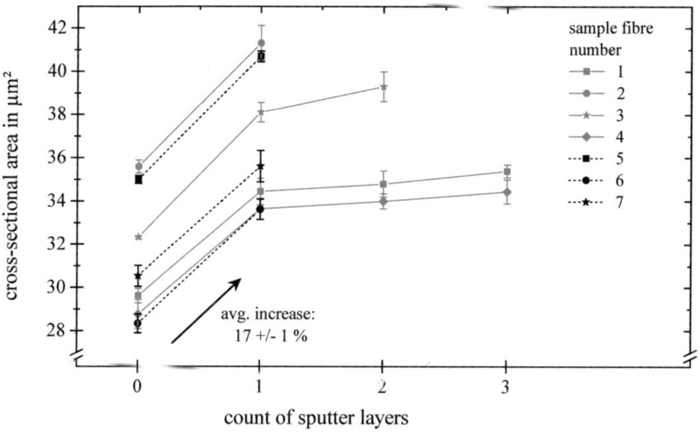

Figure 3.23: Cross-sectional area in correspondence with layers of sputtered gold: the first application of gold causes a significant increase in measured cross-sectional area while additional layers show only a slight increase.

As a consequence, it is assumed that the variation in fibre diameter is attributed to a change in the optical properties after conversion from carbon to gold on the surface. The difference in the cross-sectional area is not caused by a significant increase of the fibre diameter. Albeit possible influences of the optical properties are discussed in [163], such pronounced differences for carbon fibres are not reported in literature.

Figure 3.24 is an alternative representation of the fibres with no (0) and one sputter layer shown in Figure 3.23. Here, apparent diameters are plotted instead of the cross-sectional areas. These apparent diameters result from maximum a and minimum b of the approximation function (Figure 3.6) and are the semi-major and semi-minor axes of the ellipse, respectively. As the approximation function yields no deviation value, no error bars are given here.

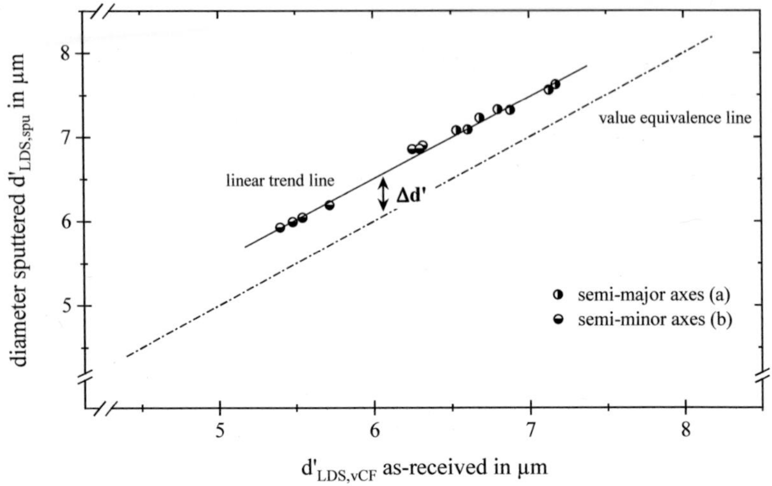

Figure 3.24: Apparent diameters calculated using the semi-major and semi-minor axes of the approximated elliptical fibre shape. Fibres with one layer of sputtered gold versus the same fibres in their as-received state. The values correspond to Figure 3.23, sputter layers 0 and 1. Cross-sectional areas are transferred to apparent diameters. The averaged difference $\Delta d'$ is 0.51 µm ± 0.051 µm.

The dashed line represents a line with slope one and all readings for gold coated carbon fibres lie above the reading of as-received fibres. On average, the fibres with gold layer yield an apparent diameter increased by 0.51 µm ± 0.051 µm, highlighted by the arrow.

Experiment 3

In the third experiment, two methods of determining the cross-sectional area were directly compared: single fibres were first measured using the diffraction sensor and the elliptical fitting procedure. Each fibre was measured at five positions at an interval of 1 mm in the centre of the fibre. Afterwards, the very same fibres were cut in the middle of the measured region and glued to a polymer bracket. The cut sections of both fibre fragments were facing upwards. Fibres and brackets were embedded in potting resin, ground and polished perpendicular to the fibre axis, removing approximately 1 mm of material from the surface. These microsections were then investigated in SEM (SEM/LEM). For analysing the cross-sectional areas, the fibre contour was traced with a polygonal and an elliptical approximation. Cross-sectional areas were calculated with an image processing software. The procedure is exemplary shown in Figure 3.25. Three polygonal and three elliptical approximations were taken and the six independent values were averaged for comparison with the LDS0200.

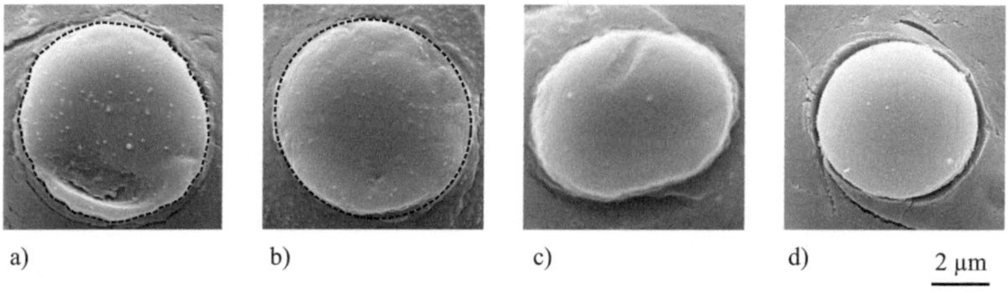

Figure 3.25: Examples of SEM images to measure the cross-sectional area of selected carbon fibres: a) vCF-1, front side, polygonal approximation; b) vCF-1, back side, elliptical approximation; c) vCF-2, kidney-shaped; d) HM-fibre. Contours of the fibres were traced and calculated with an image processing software.// SEM/IAM

This took place with a few limitations: for some fibres, only one face could be used. For others, only an elliptical or a polygonal approximation was useful. A total of 12 fibres was measured by LDS and 21 SEM images were evaluated. In addition to investigations on PANEX 35 fibres, a random sample of HM fibres was also considered. Results of this evaluation are shown in Figure 3.26. Measurements taken from the two opposite faces of the same fibre are plotted as pairs. Open symbols indicate measurements where only one side of the fibre could be evaluated.

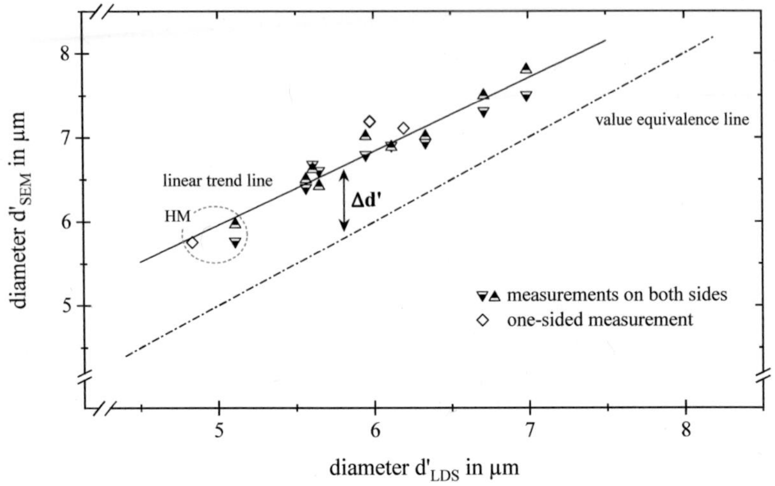

Figure 3.26: Comparison of cross-sectional areas determined by LDS and SEM//LEM. Standard deviations have been removed for the sake of clarity, but lay below 0.05 µm (LDS0200) and 0.12 µm (SEM). Encircled are results for HM fibres.

With SEM investigations, measuring inaccuracy not only result from display and resolution difficulties but can also be a consequence of user-specific assessment. Variations between two related images, that is two opposite cut faces, were up to almost 8 %. This deviation by far exceeds the standard deviations found for the six individual measurements on one SEM image of one fibre face. Standard deviation lay below 0.05 µm using the LDS0200 and below 0.12 µm for SEM. All individual cross-sectional areas determined by means of laser diffraction (LDS0200) lay below values obtained by SEM. In the aggregate, LDS values were on average only 76.7 ± 4.9 % of the cross-sectional area determined by SEM. A significant variation between both techniques was found. Despite slight deviations, the results presented in Figure 3.24 and 3.26 are consistent. Both diagrams show the same trend and indicate that the LDS0200 markedly underestimates the cross-sectional area of untreated carbon fibres.

Correcting function for the laser diffraction sensor

Given the evidence that the calibration of the diffraction sensor was optimised for thin steel wires instead of carbon fibres, a correction function was needed to determine the diameter projections d' of carbon fibres. The correction was run retrospectively to the mechanical tests. Whereas the manufacturer of the LDS0200, CERSA-MCI, only offers a standard calibration with steel wires, the company *Dia-Stron Ltd* recently designed a set of *calibration fibres* and had them gauged with high precision by applying a tactile procedure. Using this accurate calibration technique, they provide sets of fibres in the diameter range of 5 to 15 µm. A set consists of two carbon and two silicone carbide (SiC) fibres. Here, only the two carbon fibres were considered two prevent faulty results, which might occur due to material and surface differences in the fibre materials.

In Figure 3.27, idealised measurements (open star symbols) are compared to the real values provided by the manufacturer (solid symbols).

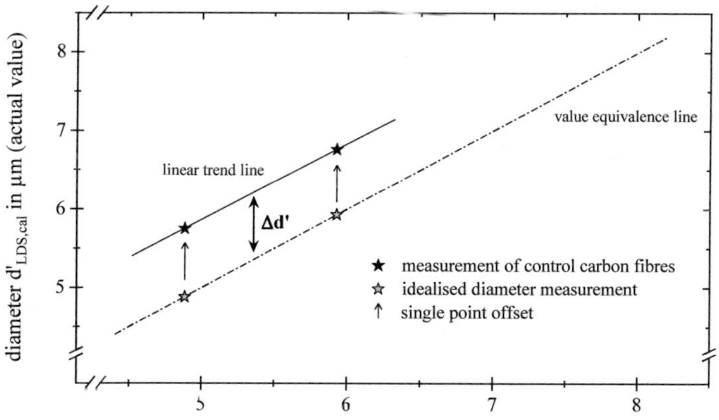

Figure 3.27: Correction of the laser diffraction sensor: The dashed line represents values found by the original calibration. The black stars are the supplier's measurements of the known *calibration fibres*. From the fit equation, a correction function was derived.

3.3 Discussion: Single fibre testing

3.3.1 Discussion of computer-aided approximation

The state of research does not offer any procedures to account for non-circular fibre shapes. The procedure introduced in this work is in the following compared to standard procedures which commonly assume circular geometries. This simplifying assumption is prone to errors even when the diameter is determined by means of laser diffraction. In this section, the virtual experiment is discussed in terms of reliability and applicability. Therefore, several interpretations of the results are shown.

Fibre tracings A to J

For all fibres A to J, the two elliptical assumptions are at least as good as the circular approaches. All circular assumptions significantly deviate from the true values. Comparing the the two best circular assumptions, *0/60/120* is superior to *0-140* due to its higher degree of symmetry. Assumption *0-140* creates an asymmetrical weighting, an effect more pronounced for elongated fibre shapes, for instance A, E and J. This effect is significantly reduced in case of the elliptical fitting procedure. *Ellipse 140* is on average as good as approach *ellipse 170*. The sinusoidal nature of the fitting function allows extrapolating to angular regions, which could not be considered experimentally. Larger deviations are also found for distinctively non-uniform fibre shapes like A, E and J; nonetheless, the results are more reliable than the circular assumption. This is for instance underlined by fibres B and G: they are both non-circular and slightly elongated and elliptical approach yields more accurate results than the circular assumption.

Geometries i to v

Fibre shapes *i* and *ii* have distinct undercuts and cause large deviations for both, the circular and the elliptical approach. It is derived that no reliable measurements by means of laser diffraction are feasible for fibres with distinct undercuts. The octagon *iii* shows slight deviations for all procedures and circular and elliptical assumptions can be considered to be equally good. A very distinct difference between circular and elliptical assumption is found for the regular ellipse *iv*. Both elliptical approaches yield results equally consistent with the true value. However, the circular assumptions deviate significantly from the true value. As expected, all approximations work very well for the circle *v*. Referring to Figure 3.19, one can see that the minimal, the maximal and the random circular assumption fail to generate reliable results.

Incremental fitting procedure for determining the cross-sectional area of carbon fibres

In order to apply the elliptical fitting procedure, at least three independent measurements in three different rotational positions must be recorded. For application of the method, it is of interest how many values are required for obtaining reliable results.

Figure 3.28 gives results for all ten fibre tracings *A* to *J* shown in Figure 3.17. The matlab script was run with different amounts of measurement inputs, starting with 3 and ending with 18 individual diameter readings. Increments of rotation in the virtual measurements were 10° and thus the same as for the real experiment. Below 20° (3 measurements), no evaluation using the script was feasible (region a)). While results vary greatly for small counts of measurements, the results stabilise above 110° (region b)). Given the geometrical constrains of the SiFiT setup, a range of 140° can be covered. Angles of more than 140° were not achievable as LDS and the frame of the testing device collide. As one can see, results vary only slightly between 140° and 170° (region c)). The term relative deviations corresponds to the derivation from the nominal value, labelled "true value" in Figure 3.16.

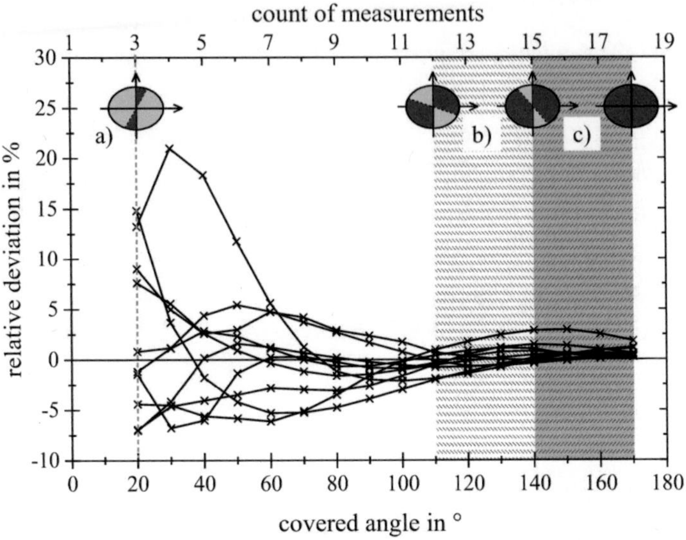

Figure 3.28: Evaluation of the quality of the elliptical approximation by calculating the relative deviation as a function of the angle covered.

Thus, the constraints of the setup do not negatively affect the measurements. The elliptical fit is qualified for approximating the real cross-sectional area as it can account for the missing data in the region from 140 to 180°. Numerical integration procedures would rely on finer steps of angular positions and are prone to errors in the region not covered by the LDS0200.

Representation of circular assumptions

As schematically shown in Figure 3.29, the calculated areas can significantly vary. The labels refer to fibres presented in Figure 3.17. The light grey tracing represent the fibre shape; the white and the dark circle are an inlying cross-section of minimal radius (r_{min}) and an outlying circle of maximal radius (r_{max}), respectively.

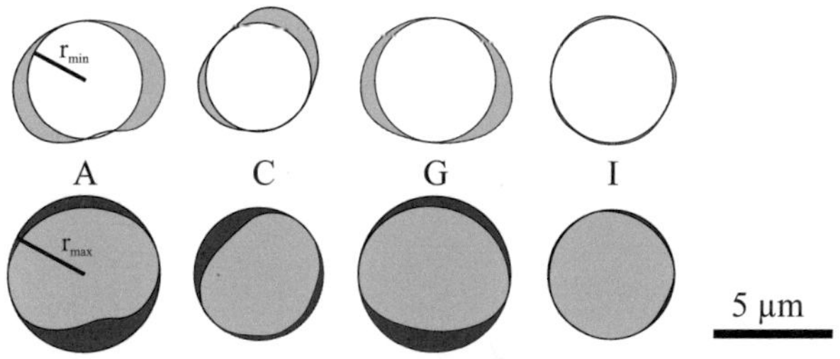

Figure 3.29: Errors when assuming circular fibre shapes. The labels A, C, G and I refer to the fibre tracings shown in Figure 3.17. The grey area is the shape of the fibre, the white circle the minimal and the dark circle the maximal circular cross-sectional area.

Figure 3.30 quantifies the findings given in Figure 3.29 for four arbitrary fibres. While variations are negligible for the fibre that is close to circular shape (I), errors are large for oval and kidney-shaped fibres (A, C, G).

In region I, only one measurement was taken. When calculating circular cross-sections the error can be decreased by individual measurements in different, regular angular positions. Thus, in region II, two individual measurements with a rotational distance of 90° were taken, in region III, three diameters in angles of 60° were recorded. As the 0°-direction would be arbitrary chosen in a real experiment, each data point in Figure 3.30 represents a different starting point; starting with 0° and rotation by increments of 10°. Interestingly, what is presented in region I of Figure 3.30 can be understood as motivation to introduce the elliptical fitting procedure.

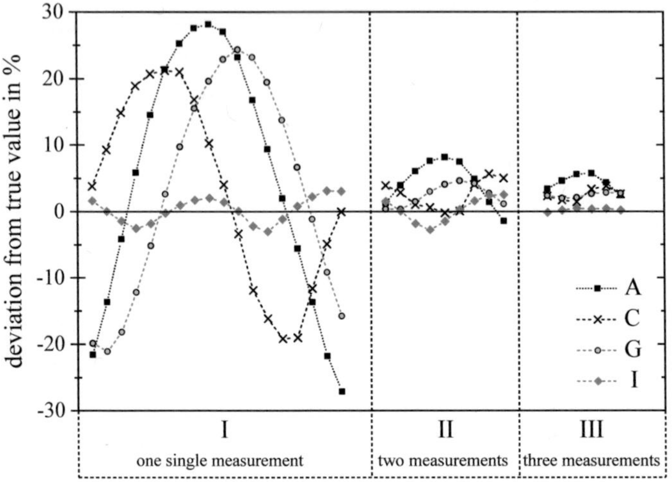

Figure 3.30: Deviation of the elliptical approximation for fibre tracings A, C, G and I in correlation to the size of the angular steps. In region I, individual results are plotted; region II and III display the average of two and three measurements, respectively.

Conclusion of the virtual experiment

In conclusion, the experiment shown here underlines the importance of a), knowing the typical fibre shape under investigations, and b), replacing circular approximations by more precise calculations.

From these results it can be deduced that the developed elliptical fitting procedure can be applied to carbon fibres of type PANEX 35, which typically possess oval shapes. Laser diffraction and the described fitting procedure are applied to all fibres investigated in the work at hand. Table 2.2 summarised the variety of different fibre shapes that can be found in literature. Given the experimental results presented here, only those of category a ("circular") in Table 2.2 can be measured by means of laser diffraction in accordance with standard ISO-11567:2018 [1]. It must be assumed that measurements of fibres of all other categories (complex, hollow, kidney-shaped and oval/elliptical) deviate significantly from the true value when applying traditional circular assumptions. For oval and slightly kidney-shaped fibres, the elliptical fitting approach enables determinations of the cross-sectional area with reduced errors. An angular increment of 10° lead to reliable results. The most important aspect is to cover a half circle in equidistant angles without doubling one direction, for instance steps of 15° from 0 to 165° [1] or steps of 36° from 0 to 144° [143].

3.3.2 Smoothening effect when applying the laser diffraction method

Representative fibre shapes are shown in the microsection in Figure 3.31. As it appears, most of the fibres fulfil the mathematical definition of convex sets given in Section 2.1.5.

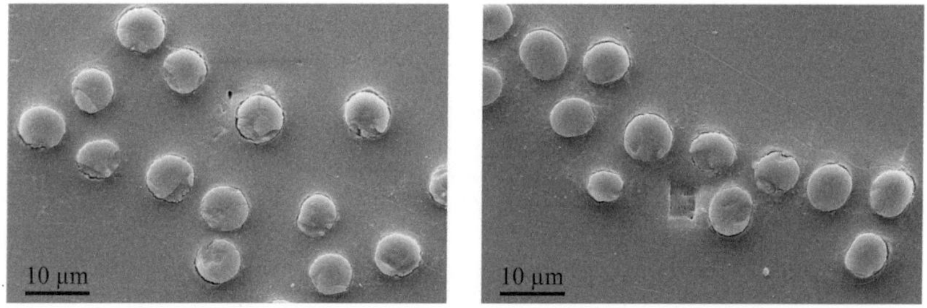

Figure 3.31: Polished faces of PANEX 35 carbon fibres investigated in SEM//IAM to demonstrate typical fibre shapes.

Nevertheless, taking a closer look at Figure 3.21, 3.25 and 3.32, suggest that the fibre surface is rough. Carbon fibres show "crenellations" and thus, strictly speaking, locally fail to meet the condition of non-convexity [33]. Laser diffraction techniques will smooth these minor dents, assuming a convex shape.

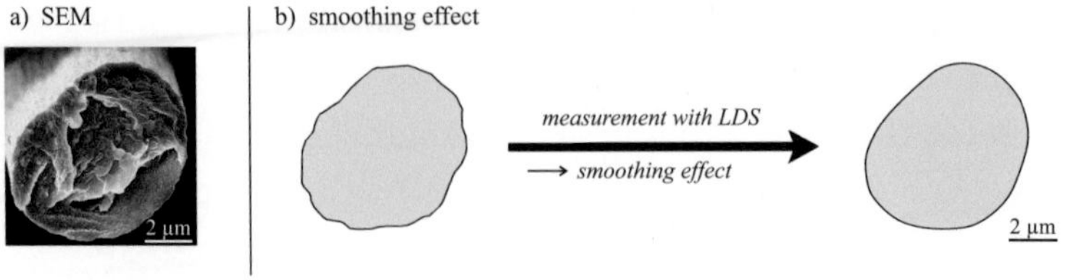

Figure 3.32: a) SEM image of the end of a carbon fibre on a fracture surface. Distinct crenellations are found on the fibre surface; SEM/LEM b) Left, a real fibre shape traced from microsection is shown. Using laser diffraction techniques to measure the cross-sectional area will inevitably lead to an unintended smoothing effect, as undercuts cannot be determined.

This "smoothening effect" can be evaluated in accordance with Figure 3.33. The exemplary image acquired by atomic force microscopy (AFM)[3] corresponds to a measurement of the circumference of a $5 \times 5\,\mu m^2$ - fibre segment (a). Debris and indentations are seen, forming the so-called "crenelated" surface described in [33]. Imagining a complete fibre circumference, the schematic, almost circular representation of a carbon fibres in (b) is derived.

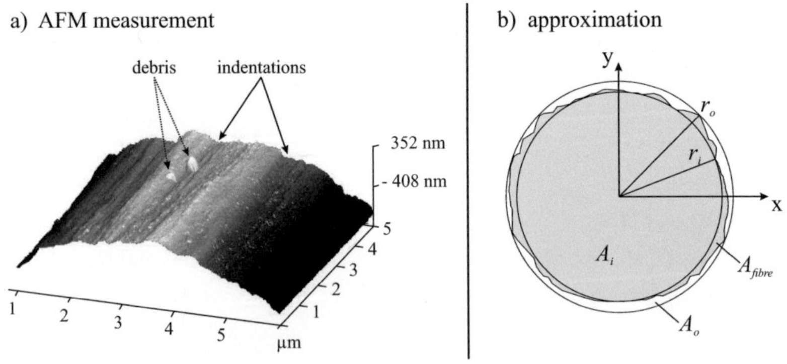

Figure 3.33: a) Exemplary AFM image of a section of a carbon fibre. Debris and crenellations are highlighted by arrows. b) Scheme of a "crenelated" carbon fibre and geometrical parameters for calculating its area.

3 For information concerning the image acquisition, the reader my refer to Section 5.1.3

According to the scheme shown in Figure 3.33 b), the following conservative estimation can be executed: again, the measuring principle of the laser is compared to a set of parallel tangents (cf. Figure 3.16), which meets the outermost edges of the rough fibre. Simplified, this correspond to the outer radius r_o and yields a cross-sectional area A_o of:

$$A_o = \pi \, r_o^2 \tag{3.7}$$

Obviously, this circular area (white plus grey) overestimates the area of the fibre (grey). Subtracting the surface roughness, the most conservative area calculation would be based on the inner radius r_i.

$$A_i = \pi \, r_i^2 \tag{3.8}$$

In this scenario, the roughness would correspond to sharp spikes rather than real fibre crenellations and it is thus assumed to yield the highest deviation from the laser measurement.

As quantitative estimation, a circular fibre shape with a mean radius of 3.5 µm is assumed. As roughness, a peak roughnesses R_{max} of 250 nm was determined by AFM measurements and is used as $r_o - r_i$ to calculate the deviation[4]. This value is within the range of other publications: to name but a few, peak roughnesses R_{max} for carbon fibres between approximately 50 and 220 nm [315–317] were reported (see also Table 7.3).

With this estimation, a deviation of up to 14 % might be caused by smoothening effects due to the laser diffraction principle. As aforementioned, this corresponds to the maximum error assuming that the crenellations have a certain height but no area. Exemplary shown in Figure 3.35, this assumption does not reflect reality in detail. The total area of the region of interest corresponds to $A_{ROI} = a_{ROI} \cdot b_{ROI}$ where a_{ROI} is the distance x_1 to x_2 and b_{ROI} equals R_{max}.

[4] The assumptions of the circular radius and the roughness are based on experimental results presented in Chapter 6, Figure 6.2 and Table 6.3.

The integral of the surface profile from x_1 to x_2, can estimated to be half the area of the region of interest. Thus, the estimated error can at least be halved to obtain a more realistic estimation of the LDS error.

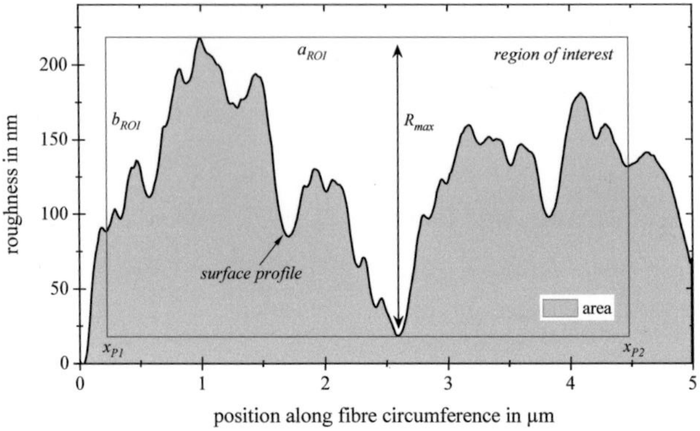

Figure 3.34: Surface profile of a 5 µm- section of a carbon fibre circumference.

One should bear in mind that the laser diffraction sensor averages over a distance of 2 mm along the fibre, by far exceeding the feature size in Figure 3.33. Summarising, the following aspects can decrease quality and reliability of determining the cross-sectional area by means of laser diffractometry. They can occur individual or combined and are also depicted in Figure 3.35, where the likelihood of erroneous readings and the complexity of corrections is raised from left to right: for a perfectly circular fibre shape (a), the limiting factor of the measurement is the accuracy of the sensor. For oval shapes (e.g. [7, 136, 151, 152]), shown in (b), circular assumptions lead to inaccurate results, which should be compensated by alternative procedures, for instance the method presented in the work at hand [110]. Several types of carbon fibres are kidney-shaped (e.g. [136, 139, 146, 154]) and do not fulfil the definition of convex sets. Laser diffraction sensor cannot assess the size of the undercut (c). As shown earlier in AFM investigations, the crenellations along the carbon fibre are another deviation from the convex shape and will cause overestimations of the cross-sectional area (d). Debris on the fibre, for instance dust and other particles, can affect the measurement as the laser sensors cannot distinguish between fibre and debris (e). In addition to these causes of error, it has been discussed in Section 3.2.2 that the calibration of the laser diffraction sensor plays an important role and that the fibre material might interact with the laser beam.

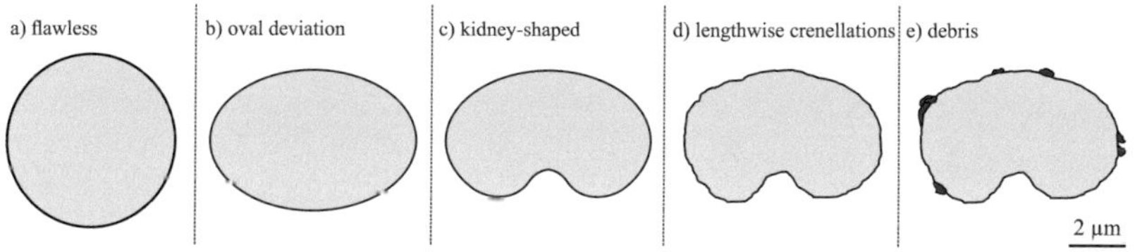

Figure 3.35: Features of the fibre which might affect the quality of the results obtained by a laser diffraction sensor.

3.3.3 Reliability of strain measurements

A comprehensive overview of fibre testing procedures was presented in Section 2.1.4. Here, the method development concerned with strain measurement on single fibres is briefly brought into the research context. For detailed remarks, the reader may refer to Section 2.1.4. Advances in digital image correlation [118, 122–124] allowed for measuring strains on fibre level.
The need for direct strain measurements is best visualised by real experimental data. Once more, the experiment is visualised (Figure 3.36, see also Figure 3.3). In the bead regions B1 and B2, a total of ten facets is selected in the digital image correlation software to yield nine pairs to form virtual extensometers. Facets for direct strain measurements are highlighted by dashed white boxes and numbers 1 to 10 in Figure 3.36. In addition, facets 11 to 16 are positioned on the gripping system. Mainly due to possible deformations within the gripping region, emphasised by the shaded boxes, measurements of direct and indirect strain measurements differ significantly. Within the gripping region, the external load is transferred into the fibre via the interface between the coated grippers and the fibre.

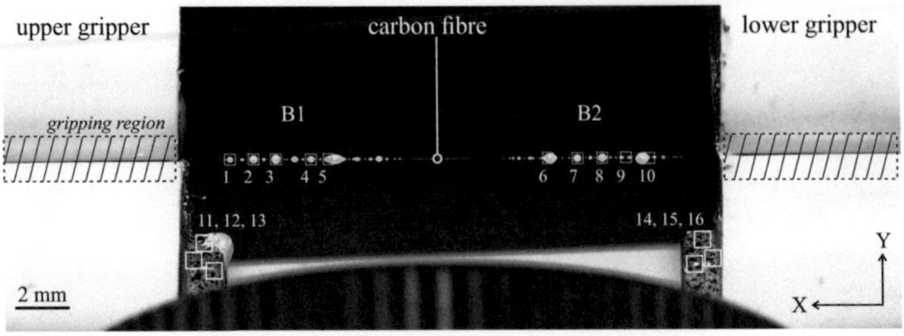

Figure 3.36: Tracking markers for direct and indirect strain measurements, positioned on the fibre and the grippers, respectively. Deformation may occur within the gripping region, where the external load is transferred into the fibre.

In Figure 3.37, optical strain measurements are plotted as a function of strain as recorded by tracking the mechanical displacement of the machine's cross head. Exemplary, results for one fibre are compared. For each strain increment determined by the testing machine, twelve pair-wise strain readings were gathered by digital image correlation according to the procedure described earlier.

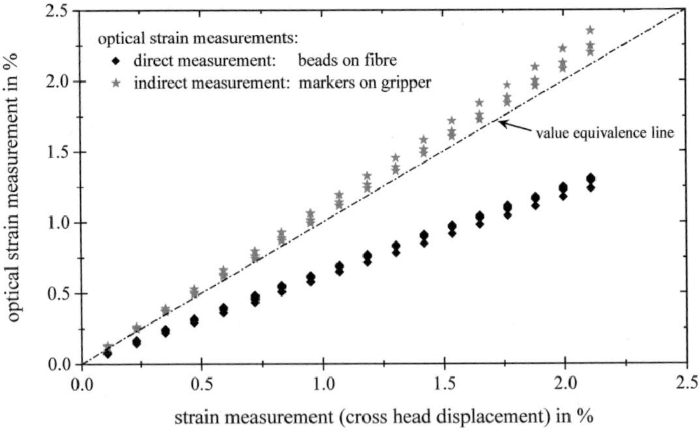

Figure 3.37: Optical strain measurements (direct and indirect) in comparison to strains obtained by the cross head displacement.

Results of optical, indirect strain measurements (on the gripper) lie rather close to the value equivalence line, which corresponds to the value provided by the cross head displacement. On contrary, the strains are significantly lower when determined by direct optical strain measurements based on the SiFiT procedure on single fibre level. Here, deviations are up to 35 %. It can be assumed that this direct (optical) measurement method better reflects reality.

Direct strain measurements are state of the art for all types of coupon size specimen [118]. Nonetheless, only few examples of direct strain measurements on fibre level were so far reported in literature: it has been run on natural fibres and thin steel wires [117, 127] wood/cellulose fibres [129, 130] and on glass and carbon fibres by the author [109, 110]. The authors were the first to use more than one tracking marker on each end of the fibre [109, 110]. Instead of flags sized several millimetres [117, 127, 128], UV-active polymer beads were attached to the fibre. Compared to paper flags and suchlike [117, 127], polymer beads are superior, as they are less susceptible to variations in lightning conditions or air draft in the laboratory. The quality of digital image correlation relies on high-resolution images, and the step-wise procedure developed here takes this fact into account, applying a 40 megapixel (350 dpi) digital camera. In this regard, it is an enhancement compared to the camera resolution for single fibre testing reported in recent publications, where the resolution was approximately 1.3 megapixel [129] or 96 dpi, respectively [117, 127].

It has been argued (cf. Figure 2.4) that direct strains measurements are needed to overcome disadvantages of system compliance corrections and slipping of fibres within the gripping regions [109, 110, 127]. The effect was investigated within this work. For all step-wise tensile test with digital image correlation, ten tracking beads were attached to the fibre and six additional markers were attached to the gripping system and evaluated pairwise (Figure 3.3). Exemplary results for a virgin carbon fibre are shown in Figure 3.38.

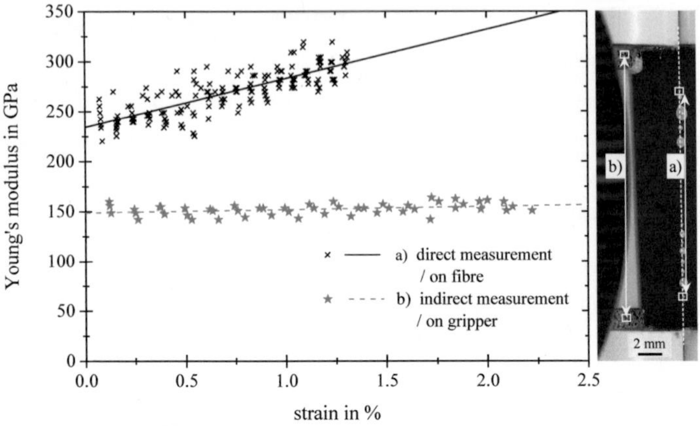

Figure 3.38: The Young's modulus calculated based on direct and indirect strain measurements. Direct strain measurements were obtained by tracking beads on the fibre (a), indirect strain measurement on DIC markers glued to the gripping system (b). The experimental data is the same as in Figure 3.37.

Here, a linear fitting procedure, as for instance proposed by Djordjević [198], is used. Two aspects are revealed: Firstly, the Young's modulus obtained by direct measurements is approximately 35 % higher than the modulus obtained by indirect measurements. Secondly, the non-Hookean behaviour of carbon fibres is not recognisable in indirect strain measurements. The approach of direct strain measurements presented by Depuydt et al. [127] found similar deviations investigating steel wires, with the direct strain measurement based on digital image correlation being approximately 25 % higher than the indirect measurement. The values obtained by direct measurement were in the range of values provided by the supplier.

It shall be noted that the grippers used in the work at hand were not optimised in terms of system compliance. Thus, the error due to indirect strain measurements might be less prone in conventional testing devices, but it cannot be circumvented. The example in Figure 3.38 underlines the importance of direct strain measurements as recommended in the standard [113].

Apart from this, quantitative comparisons with other publications have no significance due differences in setups and materials.
Independent of improvements in strain measurements, the SiFiT method allows for omitting the time consuming preparation of cardboards, the correction function to account for the system compliance and the fixation to defined measuring lenghts [111, 113]. The method presented here is expected to prove reliable for all kinds of technical and natural fibres, as the SiFiT setup can be easily adjusted to different materials [109, 110]. At the end of this chapter, Table 3.5 lists additional issues in terms of direct strain measurements in single fibre testing.

3.4 Conclusion: Test setup and its calibration

3.4.1 Calibration function

Summarising, it has been shown experimentally that the calibration of laser diffraction sensors is important. Thus it can be deduced that, before applying laser diffractometry for determining the cross-sectional area of carbon fibres, two aspects need to be evaluated. Firstly, the fibre shape of the population under investigation should be qualitatively evaluated. Secondly, the laser diffraction sensor must be adjusted to a certain material [172, 173]. Given the review of publications applying laser diffraction sensor for carbon and natural fibres, it cannot be inferred that correction calculations are already run. The literature review unveiled that correcting of irregular fibre shapes are commonly neglected, for instance when applying laser diffraction sensors on carbon fibres [179–183] or natural fibres [184–189].
For instance, factory-set standard calibration of the LDS0200 is based on steel wires. As was shown, results obtained by the factory-set calibration underestimate the real cross-sectional area up to 25 %.

To re-calculate the cross-sectional areas of the carbon fibres investigated experimentally in the work at hand, each individual diameter reading d' was corrected by Equation 3.9 and in accordance with Figure 3.27:

$$d'_c = 1.03774 + 0.96497 \cdot d' \quad (3.9)$$

The corrected values d'_c where then used as input for the elliptical fitting procedure described earlier (Section 3.1.2).

To conclude the discussion of the calibration of the laser diffraction sensor, different experiments are compared. The evaluation of a polished cross-section (Experiment 1), sputtering carbon fibres with gold (Experiment 2) and the direct comparison of LDS0200 results with top-views investigated by SEM (Experiment 3), all indicated that the laser diffraction underestimated the cross-sectional area by 15 to 25 %. The correction function, Equation 3.9, lies well within the range of the other experiments. A comparison of the four techniques is presented in Table 3.3.

Table 3.3: Experiments to evaluate the correctness of the calibration of the laser diffraction sensor: microscopy- and laser-based techniques are compared. Row "ratio" displays the ratio of the originally measured cross-sectional area A_m compared to the proposed alternative method A_{alt}.

	microscopy		**laser diffraction**	
	OM	SEM	sputtered fibres	**calibration fibres**
ratio A_m/A_{alt}	$\approx 75\,\%$	$\approx 77\,\%$	$\approx 83\,\%$	$\approx$ **78 %**
reference	Figure 3.22	Figure 3.26	Figure 3.24	Figure 3.27

The rectification with calibration fibres, highlighted in bold, were used for correcting all experimental results obtained in the work at hand. Being a direct measurement on well-known fibres, it is the most reliable calibration method.

Disadvantages of the others are: OM does not consider specific fibres but gives a general overview and SEM investigations themselves are prone to errors and are affected by the user. Sputtering the fibres impressively demonstrated (cf. Figure 3.23) that the surface material plays an important role in laser diffraction. However, it is a procedure too time-consuming to be applied prior to mechanical investigations. The calibration with calibrated fibres lies within the range of the other experiments, particularly the microscopic methods advert to a similar deviation. No matter how thoroughly the calibration procedure is executed, the "smoothening effect" of LDS causes minor measuring errors. Despite this disadvantage, one should bear in mind, that none of the other in-situ techniques offered in the standard [1] can provide remedy in this regard.

3.4.2 Concluding remarks on single fibre testing

Summary of the method development of a single fibre testing device

The state of research in single fibre testing was reviewed in Section 2.1.4. Standard procedures and approaches to measure strains directly on fibre level were presented. This chapter was dedicated to the development of a procedure to overcome disadvantages of the currently available methods. The procedure was comprehensively evaluated and it was further shown that the polymer beads used as tracking markers do not negatively influence the results.

A procedure to approximate the cross-sectional area of carbon fibres prior to mechanical testing was developed. Instead of assuming only circular shapes, a procedure to account for both, circular and elliptic fibre shapes, based on measurements of a laser diffraction sensor was introduced. A computer-based approach to ensure the quality of the method was presented. It can be concluded, that the procedure is suitable for testing real fibres. Summarising this chapter, the requirements defined at the beginning can be met with the SiFiT device. In addition, it offers the following advantages:

The SiFiT setup requires neither cardboard frames as proposed in the standards [111, 113] nor fibre preparation such as embedding, as needed for the test-

ing device LEX820/LDS0200. Just like for FAVIMAT+, only a small weight needs to be applied to the lower end before the fibre can be transferred to the test section. In terms of measuring the cross-sectional area of the fibres, the SiFiT utilises the same measuring tool (LDS0200) as presented in other publications but an advanced calculation approach. Strains are measured by digital image correlation directly on the fibre while using a step-wise loading procedure. Step-wise loading allows for high-resolution images to be taken and offers the possibility of concurrent determination of the Poisson's ratio.
To conclude, Tables 3.5 and 3.4 summarise advances in single fibre testing technology. Table 3.5 resumes challenges in regard of direct strain measurements and provides solutions how to perform reliable measurements. Table 3.4 focusses on cross-sectional areas, rating to what extent the developed approximation procedure [110] can be applied to carbon fibres and assesses alternative procedures stated in the standard [1]. In addition, it lists whether a cross-sectional area or an apparent diameter is obtained.

Table 3.4: Applicability of test methods to determine the cross-sectional areas; *Abbreviations: smo = smoothening effect; deb = debatable; d = diameter; CSA = cross sectional area.*

exp. method		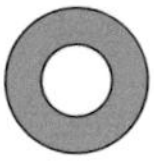			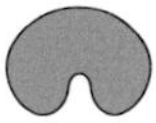	
laser diffraction	yes	no	yes	smo	no	CSA / d
	not applicable for amorphous fibres; accurate calibration needed					
vibroscopic	yes	deb	deb	deb	deb	d
	calculation method assumes circular shapes					
SEM lateral	yes	no	no	smo	no	d
	in-situ tensile test or subsequent diameter determination					
SEM top	yes	yes	yes	yes	yes	CSA / d
	not applicable in in-situ tests					
LM lateral	yes	no	no	smo	no	d
	reduced resolution					
LM top	yes	yes	yes	yes	yes	CSA / d
	reduced resolution; not applicable in in-situ tests; only embedded fibres					
LSM	*Laser scanning micrometer typically not applicable for fibres smaller* 10 µm					d

Table 3.5: Issues and solutions presented in this enhanced method for single fibre tensile tests.// *Partly published by the author in [109]*

Problem / Shortcoming	Reference	Solution / Improvement
Indirect measurement	[113, 130]	Direct measurement on fibre
Time-consuming procedure using cardboards	[113, 129]	In the SiFiT device, fibres are aligned by gravitational forces or using a Venturi nozzle
Effects of system compliance and fibre slippage	[113, 129]	Obsolete for direct measurement techniques
Sample size	[113]	Direct measurement reduces sample size as the system compliance does not have to be taken into account.
Gauge length	[113, 127]	Gauge length can be freely chosen for the SiFiT device
Fibre moves out-of-plane	[127, 128]	Out-of-plane movements are reduced by an improved alignment technique and the reduction of the bead size (markers)
Light condition	[127, 128]	A UV-light source and fluorescing beads offer more reliable image correlation as they show a distinct pattern independent of slight movements of the light source
Twisting of fibre	[127, 128]	Less relevant for spherical beads. Further, the SiFiT device can prevent twisting if the fibre is aligned using a Venturi nozzle
Influences on the fibre properties due to optical flags	[127, 128]	Replacement of the flags ($\approx$3 mm) by smaller beads ($<$ 150 µm). It was demonstrated that these do not influence the stiffness of the fibre nor do they cause fracture
Resolution	[127–130]	Using a high-resolution camera, sub-pixel digital image correlation and a step-wise procedure allow for a much higher resolution

Final remarks

To end the chapter concerned with the method development in the field of single fibre testing, some interim conclusions are given. The state of research in carbon fibre technology and single fibre testing as well as the findings presented in this section allow the following recommendations for future single fibre tests of carbon fibres. They were considered when comparing virgin and recycling fibres, Chapters 5 and 6:

- The term "diameter" is misleading in the case of carbon fibres as their cross-sectional areas often differ from perfect circles. "Cross-sectional area", "apparent diameter" or "equivalent diameter" apply much better.
- Vibroscopic measurements to determine the cross-sectional areas of carbon fibres are prone to errors [197]. Firstly, carbon fibres exhibit an anisotropic microstructure, consisting of core and skin regions that differ in their mechanical properties. Secondly, carbon fibres have cross-sections which clearly deviate from the shape of a perfect circle. Both features can affect vibroscopic measurements.
- When using laser diffraction techniques to determine the cross- sectional area of carbon fibres before mechanical investigations, it is beneficial to use procedures that are capable of approximating the real fibre shape rather than measuring apparent diameters and assuming circular shapes. Accompanying microscopic investigations can be used to determine the real fibre shapes found in a population to discuss the applicability of different procedures. Furthermore, great care has to be taken to calibrate laser diffraction sensors.
- Direct strain measurements allow to omit corrections by considering the system compliance.

4 Materials, recycling and composite production

Four bulk moulding compound materials were in the main focus of this study. To allow for comparability, they were all produced according to the same compounding process. As reference material served a hybrid resin system mixed with carbon fibres from dry fibre waste. This material was compared to composites consisting of the same type of fibres but an epoxy resin as matrix. Furthermore, materials with recycled fibres were investigated. Fibres were recycled by pyrolysis, performed at *carboNXT GmbH*, Wischhafen, Germany.

To avoid misunderstanding, Figure 4.1 presents these combinations and abbreviations used throughout this work. Lower case letters indicate the state of the fibre material; virgin fibres, in this work obtained from dry fibre cut-offs, are labelled *v* and fibres recycled by means of pyrolysis are represented by *r*. Upper case letters designate the composite materials: the resins are labelled as epoxy-based (*EP*) and hybrid resin-based (*HS*); *CF* (carbon fibre) refers to experiments with single fibres.
All materials were investigated on different levels and from microscopic to macroscopic level: for single fibres, cross-sectional areas, tensile properties and surface properties were tested. The composite interface was examined and important composite properties were investigated.

This section introduces the fibre, matrix and composite materials used in the work at hand. In addition, it presents how composite panels consisting of virgin or recycled fibres and a thermoset polymer were produced.

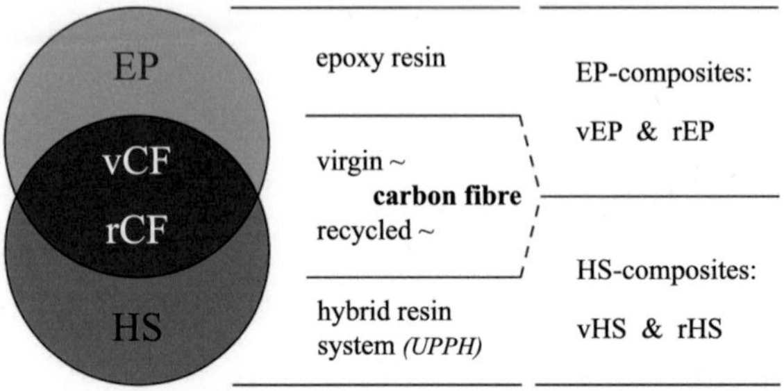

Figure 4.1: Materials investigated in this work. CF = carbon fibre / EP = epoxy resin / HS = hybrid resin / v = virgin / r = recycled

4.1 Reclaimed fibre waste

As defined beforehand and shown in Figure 2.19, recycling of carbon fibres can refer to either the reuse of non-impregnated (dry) carbon fibre fabric waste (group R-I) or the reclamation of fibres from a pre-cured (pre-preg) or cured composite (group R-II and R-III, respectively). While the dry fibre waste can be expected have retained the mechanical properties of virgin fibres, fibres that have been in contact with a resin might exhibit reduced performance. Thus, both types are compared directly within this work.

Within the framework of a scientific work, it was necessary to ensure that the original fibres and the resin are the same. As the origin of commercial fibre waste is usually unknown and unspecific, the recycling route was executed in-house according to the following procedure.
In order to allow for comprehensive comparison, as many parameters as possible were kept constant. Instead of using real composite waste, materials with a well-known history were taken into consideration. So, to simulate waste fibres, laminates were produced at *Fraunhofer ICT*[1] according to the follow-

[1] https://www.ict.fraunhofer.de - 07.05.2019

ing procedure. In a first step, carbon fibre fabric of type PANEX 35 was used to manufacture a biax composite. Eight layers of the fabric were stacked as dry preform for further processing, resulting in a preform weight of approximately 1170 g. Thermoset resin, a mix of Huntsman 3585, hardener 3458 and internal release agent in a ratio of 100 : 19 : 1.8 was prepared and heated to 80 °C. Laminates were then produced in a high-pressure resin transfer moulding process (HP-RTM), with an injection time of 60 s and a curing time of 300 s, applying a press force of 4200 kN at 100 °C. The resulting dimensions were approximately $900 \times 550 \times 3.4\,mm^3$. The matrix was then removed by means of conventional pyrolysis processes at *carboNXT GmbH* and is detailed in susequent paragraphs.

Virgin fibres (vCF)

Virgin carbon fibre fabric of type PANEX 35 by *Zoltek Toray Group* was precut on a CNC-cutting table of type *G3 L-2500* by *Zünd Systemtechnik AGX* to patches sized $50\,mm \times 50\,mm$. For laminates vHS and vEP these patches received no further treatment, particularly, the polyester stitching yarn was not removed. Here, the fibres are expected to have the same properties as virgin fibres as they have not undergone any chemical or thermal treatment [281]. The sizing was not eliminated. In state of research BMC process routes the fibre material is taken from new coils and cut just before mixing. In contrast to this, the route applied here simulates the reuse of fabric waste, contributing to group R-I recycling. Strictly speaking, these carbon fibres were not real waste. Nonetheless, to prove the applicability of recycling technologies, waste is commonly "simulated" [289]. In this way, recycling routes can be created and evaluated while ensuring comparability between virgin and recycled materials.

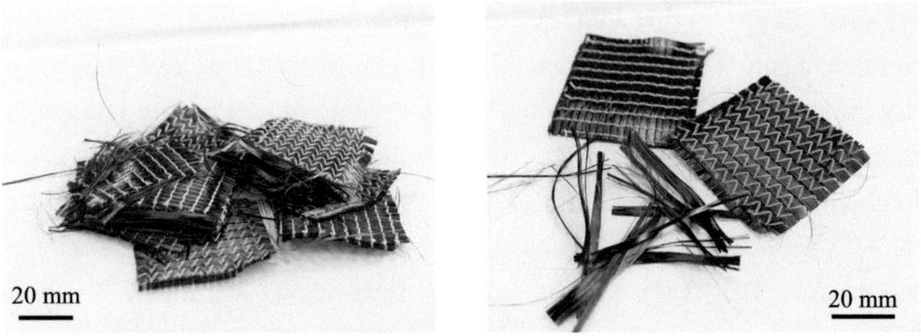

Figure 4.2: CF-patches cut to rectangular pieces of 50 × 50 mm. On the right picture, polyester weft and warp yarns on the front and the back can be seen in white. Fibre bundles could be pulled out of the fabric with ease.

Exclusively, *Zoltek Toray Group*'s PANEX 35 fibre in its virgin and recycled state were used to produce composites. To gain additional insight, two further types of fibres were investigated in single fibre experiments. Concerning the applicability of the procedure to determine the cross-sectional area and the Young's modulus, high modulus fibres, type PYROFIL HR40 by GRAFIL INC. and high tenacity fibres, T700 produced by TORAY, were investigated. Both types of fibres were taken from rovings instead of fabrics.

Recycled fibres (rCF)

In the work at hand, recycling of CFRP and carbon fibres by means of a direct BMC compounding is studied, as also presented in [150, 251]. The investigated materials were recycled at a commercial facility of *carboNXT GmbH*[2]. Thus, the recycling parameters are a corporate secret. Details concerning the parameters can neither be quoted here, nor could they be influenced. This allows for a realistic assessment of structure-property relationship of recycled CRFP. Realistic in the sense that the ambiguity about the origin of recycled material is one of the greatest challenges of composite recycling. However, it is known that the procedure is optimised in terms of a careful recovery of the

[2] https://www.carbonxt.de/de/ - 07.05.2019

valuable carbon fibres and is used in industrial-scale applications. From thermogravimetric analysis run in the experimental part, it may be assumed that the temperature was at least 430 °C, as this temperature would cause a complete decomposition of the resin [318]. Having lost their polymer stitching yarns, they were not available as patches but as a dry fibre accumulation (Figure 4.3).

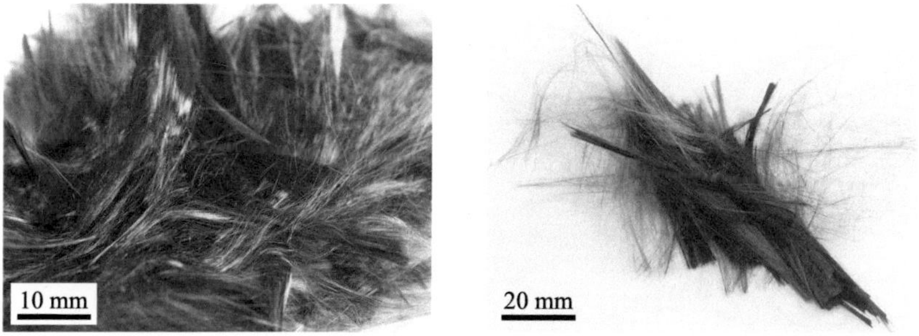

Figure 4.3: Carbon fibres recycled by pyrolysis: the fibres are still present in bundles, but show little cohesion due to the lack of sizing.

4.2 Resin materials

Throughout this work, two different resin systems were used. A hybrid resin specially designed for SMC and BMC processes, consisting of unsaturated polyester and polyurethane (UPPH, hereafter labelled HS) and an epoxy resin (EP) system tailored to meet the demands and curing profiles of BMC products. As HS was used in the majority of tests, it will be referred to as reference resin. The chemical formulation of both types of resin are listed in Table 4.1.

Table 4.1: Resin formulation HS and EP

type	supplier	description	parts by weight
hybrid resin system HS			
Daron ZW 14142	Aliancys A.G.	UPPH low viscosity	100
BYK 9076	Byk-Chemie GmbH	CF impregnation additive	3
BYK 9085	Byk-Chemie GmbH	release and flow additive	2
BYK A-530	Byk-Chemie GmbH	air release additive	0.5
pBQ 10 %	Fraunhofer ICT	inhibitor	0.3
Trigonox 117	Nouryon	peroxide	1
Lupranat M20R	BASF SE	isocyanate	24.2
epoxy resin system EP			
EP TRAC 06605	Hexion Inc.	resin low Tg	100
EP TRAC 06608	Hexion Inc.	hardener	23

4.3 Manufacturing of bulk moulding compound

In a pilot study, a variety of materials with different parameters were produced, and for some, production parameters such as mould time or mould pressure had to be varied to obtain panels with sufficient filling of the mould and good distribution of the fibre bundles. Only the optimised and standardised materials (vHS, rHS, vEP and rEP) are introduced here. In what follows, the BMC process for panel production will be explained. The BMC process in general was described in Section 2.2. All panels were produced at *Fraunhofer ICT* using a BMC kneader, produced by *Fritz Meili Zürich*. The resin mixture was blended and stirred according to the guidelines of the supplier of HS and EP and the contents of both mixtures are listed in Table 4.1. The ingredients isocyanate and peroxide were added just before the final processing of HS to prevent pre-curing reactions.

Resin and fibres were portioned by weight with respect to the desired fibre weight content of 55 wt.-%, corresponding to approximately 42 vol.-% carbon fibres. The material recipe was designed to yield three to five panels per BMC batch. The resin mix was filled into the kneader and the kneading process was started, gradually adding the weighed amounts of fibres. Fibres had to be added gradually in order to guarantee a thorough mixture of both components and a complete impregnation of the fibre bundles. The time of fibre addition was determined visually as no guidelines were available. It was aimed to knead matrix and fibres for 15 min, but according to visual inspections, the kneading time was extended if the mixing was incomplete after these 15 min.
For better understanding, the most important steps are highlighted in Figure 4.4. The BMC kneader shown in (1) was filled with the desired amount of freshly mixed resin. To that end, the chemicals were added according to the formulation in Table 4.1 (2) and stirred by an electric mixer. Prior to that, the fabric patches had been cut to $50 \times 50\,mm^2$ patches (3). The weighted amount of carbon fibre fabric was stepwise fed into the kneading machine (4). After kneading, the BMC mix was weighed, packed in gastight foil and stored at 20 °C to mature. After 6 to 8 days, the bulks were ready for being moulded. An image of an unpacked bulk of approximately 1 kg is shown in (5). The bulk was then placed in the press and moulded to the desired panel shape, here flat panels (6).

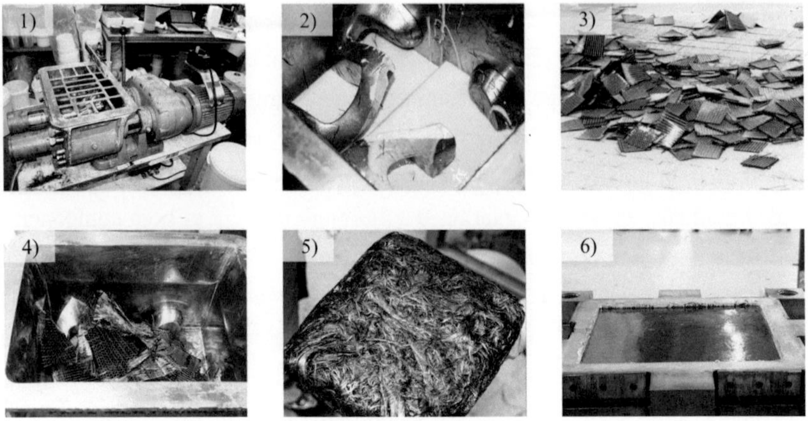

Figure 4.4: Process steps for manufacturing BMC material at Fraunhofer ICT, Pfinztal. The BMC kneader (1) was filled with resin paste (2). Pre-cut fibre fabrics (3) were gradually added (4) and kneaded to thoroughly mix fibres and resin. Subsequent, the material is removed from the kneader, weighed and formed to bulks (5). After maturing, the bulks were pressed into the desired shapes (6).

For the epoxy resin, EP, slight variations of the process were required: due to the dependency between resin viscosity and temperature, the BMC kneader had to be heated to prevent an increase of the viscosity and to improve the impregnation and mixture of EP and the fibres. A temperature control unit of type SINGLE STW-1-3-9KK2 was used to fill the heater channels of the kneader with water heated to 60 °C. Whilst the temperature distribution was uneven in the beginning, it became more homogeneous when the bulk material was kneaded. All other steps of the BMC route with EP as resin were run analogously to the HS route.

The resulting materials are displayed in Figures 4.5 and 4.6. The outer appearance of the four different materials does not vary and the materials all look alike. Neither the resin nor the state of the fibres lead to a difference in the visual appearance. The surfaces are relatively smooth but occasionally bundles disturb the flat surface.

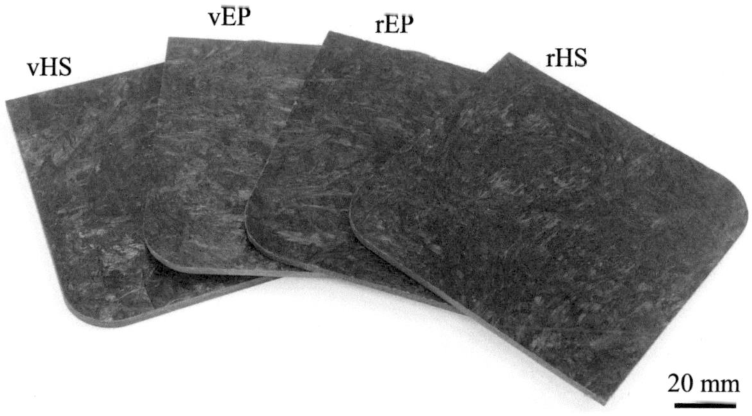

Figure 4.5: Four different BMC materials containing virgin or recycled carbon fibres and HS or EP matrix.

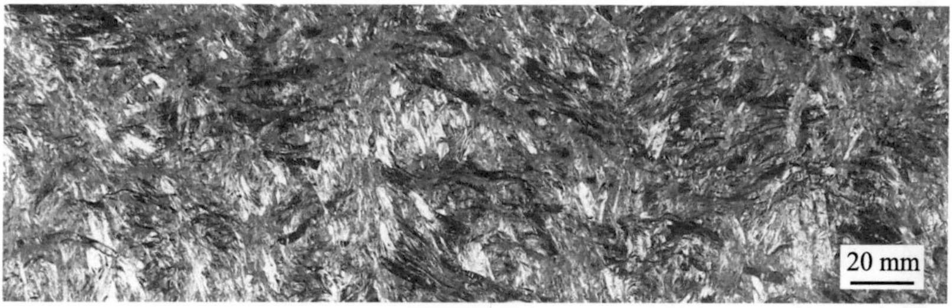

Figure 4.6: BMC reference material (vHS) in top view: the random orientation of fibre bundles can be seen.

Randomly orientated fibre bundles are visible in both illustrations. They are a general feature of BMC materials and not a result of using recycled materials. To some extent, poorly impregnated bundles or insufficiently disintegrated fabric patches were visible. Those already apparent during visual inspections of the pre-cured bulks were sorted out manually before moulding.

Summarising, two different waste fractions of the CFRP process chain were investigated: on the one hand, dry fibre fabric waste, considered to contain the original fibres with virgin fibre properties. On the other hand, composites con-

taining recycled fibres reclaimed by pyrolysis. Thus, two out of three waste fractions categorised in Figure 2.19, namely R-I and R-III, are investigated in the work at hand. In addition, two different resins were evaluated, with HS, a hybrid resin system serving as reference thermoset.
Information about the material was gathered in terms of mechanical and surface properties and is presented and discussed in the following chapters.

5 Experiments

This chapter presents the details of all characterisation methods used in the work at hand. It includes different scales, starting on micro-level, for instance scanning electron microscopy or single fibre testing and ending on macro-level by explaining the test methods for composites.

5.1 Microscopic investigations

5.1.1 Evaluation of fibre shapes by light microscopy

Light microscopy was used to evaluate the shape of embedded carbon fibres in accordance with standard ISO-11567 [1] and to study the microstructure of BMC specimens. Solely virgin fibres type PANEX 35 and their composites were examined and no quantitative analysis was run. Microsections were prepared by grinding with abrasive papers of grades P600, P1000, P2500 and P4000. For evaluating the fibre shapes, two types of specimens were prepared: one test sample was a randomly chosen cross-section of a unidirectionally reinforced carbon fibre polymer with fibres PANEX 35 and HS resin, labelled sample *CFRP*. The second was a bundle of fibres PANEX 35 (vCF) embedded in potting resin (labelled *CF*). In addition to the described grinding procedure, specimen *CFRP* was polished for 5 min with a diamond suspension containing particles sized 5 µm. Images were taken rectangular to the fibre face with a light microscope. Results are shown in Figure 6.3. In addition, light microscopy was used to investigate the microstructure of BMC specimens and to draw conclusions about the fibre impregnation. These specimens were cut perpendicular to the top face of the BMC panel. Results are displayed in Figure 6.28.

5.1.2 Scanning electron microscopy for investigating fracture surfaces and carbon fibres

Scanning electron microscopy (SEM) was used to investigate fibres and composites. Three different SEM devices have been used at Karlsruhe Institute of Technology and the University of Vienna. The devices are listed and labelled in Table 5.1.

Table 5.1: Scanning electron microscopes used in the work at hand. All SEM images were acquired by secondary electron imaging. In addition, typical ranges of working distance (WD) and acceleration voltage (EHT) used in the experiments are listed.

university	SEM	label	operation parameters	
			WD in mm	EHT in kV
Karlsruhe Institute of Technology Institute for Applied Materials	ZEISS EVO 50	SEM/IAM	6.5 - 13.0	7.5 - 15.0
Karlsruhe Institute of Technology Laboratory for Electron Microscopy	ZEISS LEO 1530	SEM/LEM	5.0 - 8.3	5.0
University of Vienna Faculty of Physics	ZEISS Supra 55 VP	SEM/VIE	7.1 - 7.5	2.0 - 5.0

In terms of single fibres, SEM was carried out to investigate the fibre surface and to draw conclusions whether recycling had any influence. In terms of composites, fractured surfaces of tensile test specimen were investigated to evaluate failure mechanisms.

5.1.3 Surface roughness investigations by atomic force microscopy

Using atomic force microscopy (AFM), influences of the pyrolysis on the surface appearance of virgin and recycled fibres were investigated. AFM was run at the University of Augsburg, in accordance with the standard tapping-mode procedure used at the *Institute of Physics / Experimental Physics II*, for instance published in [319, 320]. AFM was run on a *Bruker Dimension ICON* system with a standard tip and in tapping mode. Images were processed in the software *Nanoscope Analysis 1.5*.

5.2 Single fibre testing procedure

An enhanced single fibre testing method was detailed in Chapter 3. The method development focussed on novel procedures to determine the cross-sectional areas of carbon fibres and to measure strains directly on single fibre level. In the following, only some key facts concerning the specific test execution are presented.

5.2.1 Determination of cross-sectional areas of carbon fibres

Prior to mechanical testing, the cross-sectional area of the carbon fibres was determined according to the procedure developed within the frame of this work and elucidated in Section 3.1.2. In summary, this means that 15 measurements of the fibre diameter were taken in angular increments of 10° while a pretension of 5 mN was applied. Using the elliptical fitting procedure, the cross-sectional area was calculated.

The results obtained prior to tensile test were not only used for stress calculation but also as comparison to results acquired by scanning electron microscopy. In SEM studies, the fibres were investigated perpendicular to their lengthwise direction as this is common practise [1]. To allow for comparison with the LDS measurements, the measured diameter d was transferred to a cross-sectional area by assuming a perfect circle with the area being:

$$A = \left(\frac{d}{2}\right)^2 \cdot \pi \tag{5.1}$$

SEM measurements were taken at SEM/IAM and SEM/UVI and the following protocol was used: fibre bundles were attached to sticky tape and placed in the SEM chamber. Spots were chosen randomly and the focus of the image was set to the edges of the fibre. Different magnifications (1,000× to 10,000×) were recorded. Subsequently and as represented in Figure 6.1, measurements were taken in the software *ImageJ*, using the SEM metadata for correct scaling of the images. For obvious reasons, fibres were either investigated by the LDS0200 or SEM. Results of SEM and LDS measurements are presented in Figure 6.2.

5.2.2 Determination of strain and Young's modulus

Experiments to determine the Young's modulus were uncoupled from experiments to measure the tensile strength. Young's modulus was obtained following the step-wise loading procedure defined in Sections 3.1.3 and 3.1.4. As experiments had been run before applying the correction function of the laser diffraction sensor (cf. Section 3.2.2), the original setting was based on a cross-sectional area calculation relying on the factory-set calibration with steel wires. Later, a re-calibration of stress levels to account for the optical properties of carbon fibres was run according to Equation 3.9. Therefore, the set stress increments after re-calibration were about 195 ± 5 MPa for the tested high tenacity fibres and 179 ± 2 MPa for the high modulus fibres.

5.2.3 Tensile strength of carbon fibres

Gauge lengths were chosen to be 1, 5, 10, 20, 50 and 70 mm in order to cover a wide range of fibre lengths. As recycled fibres were only available in a total length < 50 mm and as approximately 15 mm of free length on both ends are needed to handle the fibres in the SiFiT device, the recycled fibres could not be tested with a gauge length larger than 20 mm (cf. Figure 3.2).
Carbon fibres have statistically distributed flaws, and thus, tensile strength is commonly determined following the Weibull procedure, by way of example applied in [136, 137]. The evaluation was carried out as follows: an amount

of n fibres was tested. Tensile strengths were listed in ascending order and numbered *1,2,...,n*. Each tensile strength is connected to a failure probability F_i, given by:

$$F_i = \frac{i-0.5}{n} \tag{5.2}$$

The Weibull plot is then given as $ln\ ln\ (1/(1-F_i))$ versus $ln\ \sigma_i$. There, the slope m defines the Weibull modulus m, with higher values corresponding to lower scatter.

Statistical estimation to determine an appropriate sample size

A statistical estimation procedure was run prior to the main experiment to calculate the sample size needed to obtain reliable results for the Weibull strength of carbon fibres. Two opposing goals were to achieve: a high statistical certainty while minimising testing time and expenditure [321].
In order to use Weibull's statistics when evaluating the tensile strength of high-performance ceramics, standard DIN EN 843-5 recommends a sample size of at least 30 specimens [321]. In contrast, Thomason summarised that, in case of testing glass fibres, at least 80 individual fibres should be examined when applying Weibull's statistics on single fibre testing results [220]. As a compromise, it was assumed that 50 individual tensile tests were a data base sufficiently large for the statistical estimation shown here.
The experiment was run on a preliminary version of the SiFiT. In this particular case, cross-sectional areas were calculated assuming a circular shape, based on four individual diameter measurements in angular increments of 45°. For the experiment, fibres of type PANEX 35 were tested to failure with a testing speed of 1 mm/ min and at a gauge length of 50 mm. Fifty valid measurements of tensile strength were used as a base for a statistical urn model. Firstly, Weibull's strength (σ_W) of the 50 fibres was determined. Secondly, the 50 values were defined as the population P. From the population, n_s different values were picked according to an urn model without multiple drawing.

For each n_s, the Weibull strength was calculated and the virtual experiment was reiterated a 100 times. This parallels an experiment with n_s single fibres, which is repeated a 100 times. The average and the standard deviation for each set of 100 measurements is calculated. The variable n_s was run from 5 to 50, with the latter characterising the real measurement.
The calculated averages of the Weibull strength of 100 virtual tensile tests are plotted in Figure 5.1 as Weibull strengths $\pm$ standard deviation, normalised to the Weibull strength experimentally determined for n = 50.

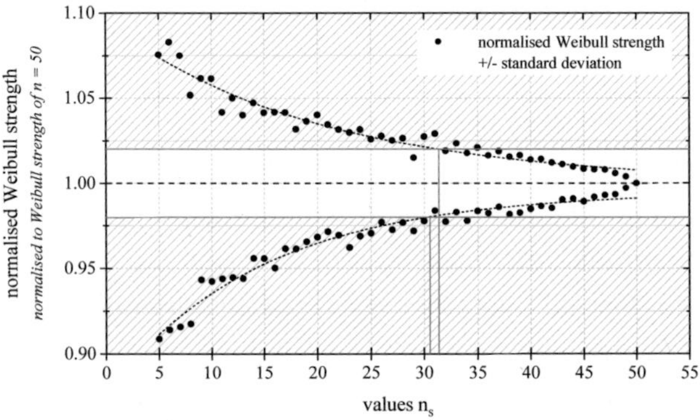

Figure 5.1: Average Weibull strength $\pm$ standard deviation, normalised to Weibull strength of sample size n = 50, plotted as a function of values n_s out of pool *P*. The non-shaded rectangle corresponds to the 98% confidence interval.

The higher n_s, the lower the standard deviation of the computed experiments, corresponding to a higher statistical certainty. In Figure 5.1, a confidence interval of 98% was selected and is highlighted by the non-shaded rectangle. To stay within this confidence interval, the results suggest that it is advisable to test 32 or more fibres. As a consequence, no less than 32 fibres were tested for each gauge length and state (virgin or recycled) of the fibre.

5.2.4 Poisson's ratio of carbon fibres

For determining Poisson's ratio of single carbon fibres, the procedure explained in section 3.1.4 was executed. The set nominal stress increment $\Delta\sigma$ was approximately 390 ± 10 MPa. The free length between the grippers was 50 mm and thus only virgin fibres were investigated.

5.3 Surface characterisation of carbon fibres

In cooperation with the PaCE[1] group at the University of Vienna, several experiments concerning the chemical and physical properties of the carbon fibres were carried out. The question was whether the fibre surface chemistry was affected by the recycling procedure. In general, it has to be expected that the sizing of the virgin fibres is eliminated due to the pyrolysis conditions. This might affect the bonding of the resin to the fibre and therefore influence interface properties.

5.3.1 Determination of surface energy: Inverse gas chromatography

By means of inverse gas chromatography (iGC), the surface energy of vCF and rCF were measured at 30 °C and a relative humidity of 0 %. Fibre bundle samples were placed in a measurement column with an inner diameter of 4 mm. A *Surface Energy Analyser* (*SMS: Surface Measurement Systems*) was used. Octane retention at various coverages was used to determine the specific surface area of both materials. The specific surface area was calculated by applying the BET model, using the maximum peaks [266].
For determining the surface energy, a series of unpolar alkanes (hexane, heptane, octane, nonane and decane) as well as polar probes (dichloromethane,

[1] Polymer and Composites Engineering - University of Vienna - https://mc.univie.ac.at/pace/ - 07.05.2019

acetone, acetonitrile and ethyl acetate) in He-carrier gas was passed over the columns containing carbon fibres. Retention times and volumes were recorded. The retention times and coverages of the organic solvents were used to calculate the dispersive surface energy according to the Dorris-and-Gray model.

5.3.2 Dynamical vapour sorption

Carbon fibres are generally believed to be hydrophob. However, dynamic vapour sorption was used to investigate whether the recycled fibres absorp more water than virgin fibres do. Test were run with a dynamic gravimetric water sorption analyser (*DVS Intrinsic*, *SMS: Surface Measurement Systems*, Alperton-London, UK). Samples of approximately 17 mg were placed in sample pans. Pre-conditioning took place at 0 % relative humidity for 24 hours at ambient temperature. After conditioning, relative humidity was set to 95 % and the samples rested in the device for another 24 hours. The water uptake was defined as the difference between the minimum of the desorption phase and the maximum of the absorption phase. The experiment was run twice with the same specimen.

5.4 Characterisation of the fibre/matrix-interface

The interface properties τ_{app} (apparent interfacial shear strength), τ_d (local interfacial shear strength), τ_f (interfacial frictional stress) and G_{ic} (critical interfacial energy release rate) were calculated based on procedures published in [252, 322]. Pull-out and push-out properties were examined according to the following experimental procedures.

5.4.1 Single fibre pull-out test

Single fibre pull-out tests were run at the *Faserinstitut Bremen e.V.* (FIBRE), according to a procedure described in detail in [322]. As two-component epoxy resin, *Epikote Resin Trac 06150* was combined with hardener *Epikure Curing*

Argent Trac 06150 was used. Fibres were taken from bundles of vCF and rCF. The load cells capacity was 2.1 N and fibres were pulled with a test speed of 0.1 mm min^{-1}. Determination of the cross-sectional area of the fibres was not attainable. The apparent diameter was set to 6.27 mm.

5.4.2 Single fibre push-out test

In contrast to microbond and pull-out test, push-out test have to be run on thin sections of composite specimens instead of polymer droplets attached on single fibres [252]. Push-out specimens were prepared for materials containing the reference resin, that are vHS and rHS. Out of the composite panels, slices of thickness $t \approx 1.5$ mm were cut and attached to a specimen holder. Specimens were then ground with abrasive papers of grades P600, P1000, P2500 and P4000 to reach a slice thickness between 50 and 70 µm. Subsequently, they were polished with a 3 µm suspension to improve the surface quality. Further processing took place at the University of Augsburg, *Institute of Physics - Experimental Physics II*. In additional lapping steps the thickness was decreased below 40 µm and the test-side of the specimen was polishing with a solution containing SiC particles sized 3 µm.
Mechanical tests were run on an ASMEC *Universal Nanomechanical Tester* (UNAT). As indenter, a flat cone diamond tip with a diameter of 4.9 µm was used. Push-out tests were run in continuous mode.

5.5 Composite characterisation

Composites were characterised in terms of their physical properties (density and fibre weight content) and mechanical properties (tensile test, dynamic mechanical analysis). Procedures and setups are described in the following.

5.5.1 Thermogravimetric analysis

Thermogravimetric analysis (TGA) was run at Fraunhofer ICT and used to evaluate if panels investigated mechanically fulfilled the desired fibre weight content. Specimens were either round discs with a diameter of 25 mm or squares of 20 $\times$ 20 mm. Specimens for TGA were taken at arbitrary positions within the composite panels. The same panels were used for TGA investigations and mechanical tests. Additionally, variations of the fibre weight content due to an uneven distribution of fibres was investigated. To do so, 64 TGA specimens were cut from one reference (vHS) panel. Specimens were positioned as a 8$\times$8 - grid of equidistant discs. To obtain the fibre weight content, an experimental analysis according to Bücheler et al. [318] was run. The time-temperature-cycle depicted there was an optimised cycle for the hybrid resin system HS [318]. The procedure was followed without any additional amendments for both types of resin.

5.5.2 Density of carbon fibres

Density of fibres and composites were measured and the results are displayed in Table 6.6. Measurements were performed at the University of Vienna using gas pycnometry in accordance with [323]. The applicability of gas pycnometry to investigate carbon fibres was researched by Rude et al. [324]. Density of composites and fibres was measured using a gas pycnometer of type *AccuPyc II 1340* (*Micromeritics GmbH*) and using helium as gaseous displacement medium. The machine was set to run ten individual experiments on each sample. Composites and fibres were prepared as follows:

For investigations on composites, ten rectangular pieces of approximately $9 \times 9\,\text{mm}^2$ were cut with a diamond saw. Samples were taken from random positions from the panels that had also been used for the mechanical tests. The pieces were given random numbers and per experimental run, two pieces were picked and measured at the same time to improve accuracy by using a higher sample weight and to account for variations in the fibre-to-resin ratio of

the composite panels. Prior to testing, the samples were stored in an oven at 70 °C for at least twelve hours.
Fibre bundles were manually cut to lengths of approximately 2 mm or 5 mm using a scalpel. In addition, *CFF*, a filler made of cut carbon fibres, was measured. Cut fibres and the filler were dried in an oven at 100 °C for at least twelve hours. At least 15 min before weighing the sample, the material was removed from the oven and sealed in a plastic bag in order to cool without absorbing moisture. Samples were weighed and placed in the sample holder of the gas pycnometer. For each sample, ten separate experiments were run automatically.

5.5.3 Dynamic mechanical analysis

Dynamic mechanical analysis (DMA) is a common tool to determine the glass transition temperature T_g of polymers [325]. DMA was executed in three-point-bending mode according to ASTM standard D5023-15 [326] for all four material combinations. A step-wise temperature sweep procedure was followed. As discussed in [325] and in ASTM standard E1640-18 [327], different evaluation procedures are available. Here, the data were analysed as follows: the temperature related to the maximum of the cubic spline fit function of the loss factor was chosen as the glass transition temperature T_g.
The experiments were run on an *Instron E3000 electro puls* testing device equipped with a 5 kN-load cell and a temperature chamber. Standard specimen size was 125 mm × 15 mm × 3 mm. The temperature range was approximately 30 to 160 °C. In increments of 2.5 °C, the temperature was held for a dwell time t_{dwell} of 120 s before the specimen was loaded 20 times at a frequency of 1 Hz. Cycles 1 to 10 were discarded and the mean value of cycles 11 to 20 was calculated and recorded. The supporting bars had a distance L_s of 80 mm; the

mean deflection was 0.3 mm and the amplitude ± 0.1 mm. According to DIN EN ISO 14125, the strain in the outer fibre is then calculated by:

$$\varepsilon = \frac{600 \cdot s \cdot t}{L_s^2} \tag{5.3}$$

with s being the deflection of the central point and h the thickness of the specimen. The mean deflection chosen in the experiment corresponds to a strain of 0.084 %, the strain amplitude to 0.028 %, respectively. Three specimens were tested for each material and were taken from those panels also considered for tensile tests.

5.5.4 Tensile testing of composite materials

Two alternative modifications of tensile tests were executed. Specimen geometries are displayed in Figure 5.2. It shows the specimen geometries and the range of validity. At what follows, both alternative procedures are presented.

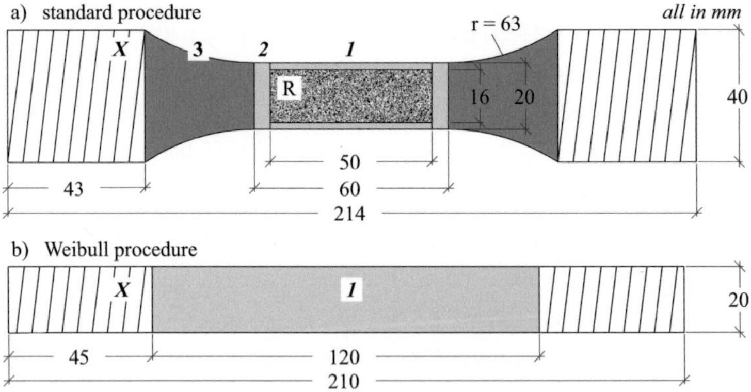

Figure 5.2: Geometry and region of validity for tensile specimens; standard procedure with dog-bone shaped specimens (a) and Weibull procedure with rectangular specimens (b). In a), region R was used for DIC. For calculating stiffness, specimens failing in region 1, 2 or 3 were considered; for determining tensile strength, specimens failing in region 1 and 2 were analysed. For specimens of type b), stiffness and tensile strength were calculated for specimens where failure occurred in region 1. For both types, specimen evaluation was annulled if cracks occurred in region X.

Approach 1: standard procedure

Tensile tests were run according to standard DIN EN ISO 527-1 [328] and DIN EN ISO 527-4 [329] with dog-bone-shaped specimens cut from BMC panels as shown in Figure 5.2. A universal testing device type *Zwick//Roell Zmart.pro 200kN* equipped with a 200 kN load cell and hydraulic grippers (20 bar) was used. Specimens were gripped with a gauge length of 128 mm. Strains were measured by means of digital image correlation using the system ARAMIS 4M by GOM. The recording rate was 2 Hz. In accordance with the standard, a centred rectangle of 60 mm length and 16 mm width was set as the region of interest (ROI) in the DIC-software. A width of 16 mm was chosen to eliminate edge effects (cf. Figure 5.2). The initial load was 20 N and the testing speed 2 $mm\,s^{-1}$. For general investigations, the strain in tensile direction was averaged within the ROI. Due to the brittle behaviour of BMC and deviant from the standard, a range of 0.05 % to 0.15 % was chosen for determining the Young's modulus.

The setup is shown in the images in Figure 5.3: a) shows the combination of the universal testing machine and the GOM DIC system, b) displays the back of a BMC specimen at the moment of fracture, apparent from spallings on the surface and c) is the front view with the characteristic speckle pattern. The specimen (S) is aligned by the help of brackets (B) and clamped with hydraulic grippers (H). Two lamps emitting blue light (L) are used to uniformly illuminate the specimen and two cameras (C) record the experiment. Within the region of interest (R), strains are evaluated. The test rig is labelled as (T).

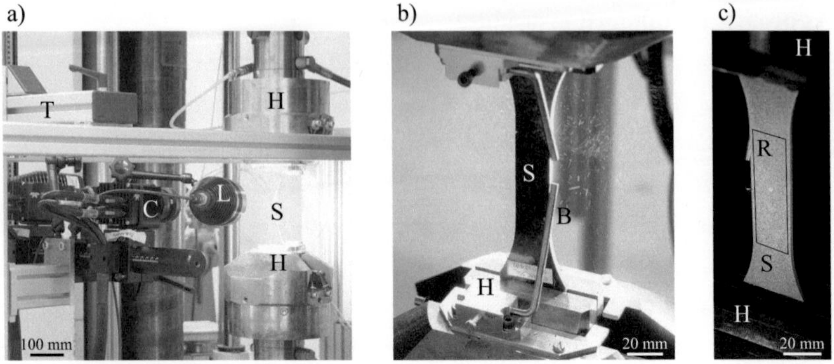

Figure 5.3: Setup for tensile tests: a) is a view upon the tensile test setup including the DIC system; b) shows the back of a specimen at the moment of fracture, c) front of specimen with DIC speckle pattern and the region of interest.
Labels are: B) brackets for alignment; C) camera system; H) hydraulic clamps; L) blue light; R) region of interest; S) specimen, T) test rig.

Approach 2: Weibull procedure

For the alternative tensile testing, rectangular specimens (Figure 5.2) were investigated. The same testing device and hydraulic clamps (Figure 5.3 b) as for approach 1 were used and an additional load cell with a capacity of 20 kN was installed. Here, strains were measured with an extensometer (Zwick//Roell multiXtens) instead of digital image correlation to enable a higher throughput and the increased number of samples needed for a Weibull analysis. The pins of the extensometer had a distance of 50 mm and where placed centrally in respect to the measuring length. When a pretension of 20 N was reached, the pins of the extensometer were automatically attached to the sample. Tests were run with a speed of 2 mm min^{-1} and ended when a load drop of 80 % occurred.

6 Results: Recycling induced properties of carbon fibres and their BMC composites

Experimental procedures were introduced in Chapter 5 and here, the results are presented. Again, investigations start on micro and fibre level and end on composite scale. In most cases, virgin and recycled materials are compared. However, when the need arises, they are complemented by general statements on fibres and composites.

6.1 Results of single fibre tests

Single fibres were characterised with respect to their physical, chemical and mechanical properties and results are presented hereinafter.

6.1.1 Assessment of cross-sectional areas of carbon fibres

One of the goals pursued in this work was to improve the measurement of the cross-sectional area of carbon fibres prior to mechanical testing. In light of the fact that mechanical properties vary as a function of the cross-sectional area [24, 58, 223], this chapter starts with the results of the latter. It deals with both, a quantitative and a qualitative analysis of the cross-sectional areas.

Quantitative results of cross-sectional areas

Investigations by SEM and LDS are compared in the following: in scanning electron micrographs, five diameter measurements were taken approximately equidistantly. The line width was set to 150 pixels to ease finding a line orthogonal to both fibre edges, as symbolised by the semitransparent box in Figure 6.1 a), diameter number 2. Figure 6.1 b) highlights the variations of PANEX 35 carbon fibres. Images were recorded with SEM/UVI and SEM/IAM at magnifications ranging from 1,000× to 10,000×.

The result of both procedures to measure the cross-sectional areas of carbon fibres are compared in Figure 6.2, where LDS and SEM measurements are compared as box plots.

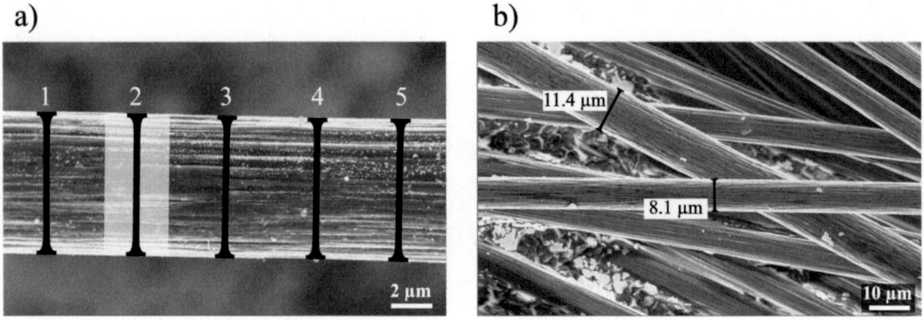

Figure 6.1: Diameter measurements on SEM images: a) Five individual measurements were taken on each image in magnifications from 1,000× to 10,000×, here exemplary shown for 5,000×. b) Qualitative illustration of the broad range of fibre diameters found in PANEX 35. Measurements emphasising this scattering range are shown in Figure 6.2. // *Images: SEM/UVI.*

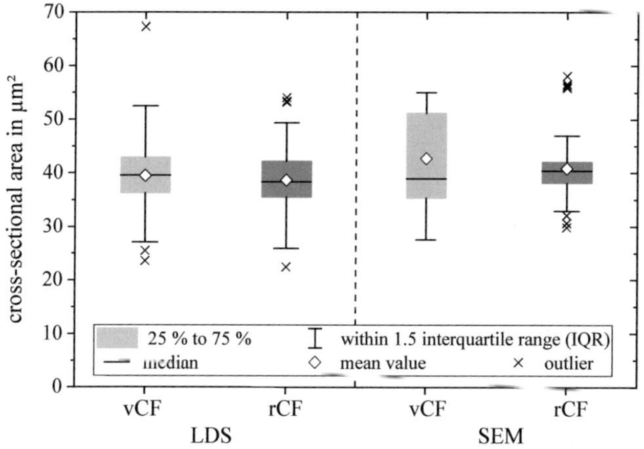

Figure 6.2: Comparison between LDS and SEM to measure the cross-sectional area of carbon fibres. Columns 'LDS' and 'SEM' reveal that neither a significant decrease of the cross-sectional area due to recycling, nor a distinct difference between both measuring techniques is found. Column 'SEM' includes results measured on side view images taken with SEM/UVI and SEM/IAM. To allow for comparison, the diameter readings taken by SEM were transferred to circular cross-sectional areas.

To obtain results by LDS, 15 separate diameter readings were taken stepwise following the procedure explained earlier for each single fibre. In this experiment, the same cross-sectional areas examined prior to tensile testing (cf. Table 6.2) are taken into account. LDS and SEM show high consistency, but the mean value of the cross-sectional areas determined by SEM yields slightly larger values compared to LDS. Independent of the procedure, the mean values of the cross-sectional areas of recycled fibres are slightly but not significantly smaller than that of vCF. For all materials but vCF determined by SEM, median and mean value are almost equal. Here, the median of the cross-sectional area of rCF (SEM) is increased compared to vCF. The broad variation of measured values is well comprehensible in view of Figure 6.1 b): PANEX 35 fibres tend to show a wide range of fibre diameters.

Qualitative evaluation of the fibre shape

In Section 3.2.2, the quality of the elliptical fitting procedure has been assessed on base of fifteen representations derived from microsections of carbon fibres as well as regular geometrical bodies. In addition to this general and computer-based quality assessment, experimental results on PANEX 35 are presented in the following by categorising the cross-sectional areas with increased statistical certainty. In this evaluation, the focus lies upon fibre shapes rather than on a qualitative measurement of the cross-sectional area.
Verification was run on two different test samples prepared according to Section 5.1.1. A total of 127 fibres were part of sample *CFRP*, and 436 fibres were investigated for sample *CF*. Fibres were visually categorised, designating them to one of the following four groups:

- **Convex, smooth:** This category strictly fulfils the definition given for convex sets in Section 2.1.5. The fringes of these fibres are smooth.
- **Convex, rough:** The fringes of these fibres are rough and show some irregularities. Picking two points close to the fringes, the definition of convex sets is not fulfilled. However, it is assumed that these small deviations will have a minor effect as the superordinate structure is still convex, see Figure 3.32.
- **Non-convex, rough:** These fibres do not fulfil the definition of convexity, stated in Section 2.1.5. In addition to a rough surface, their shape approaches a kidney-shape.
- **Non-convex, smooth:** These fibres are slightly kidney-shaped while having smooth fringes. At some geometric points, the requirement for convex-sets is not strictly fulfilled.

The results are displayed in Figure 6.3. The majority of fibres (89.4 %) shows a convex shape. The remaining share expresses only a minor tendency of being kidney-shaped (non-convex) and thus the procedure explained in Section 3.1.2 was used to determine the cross-sectional area prior to mechanical testing of carbon fibres.

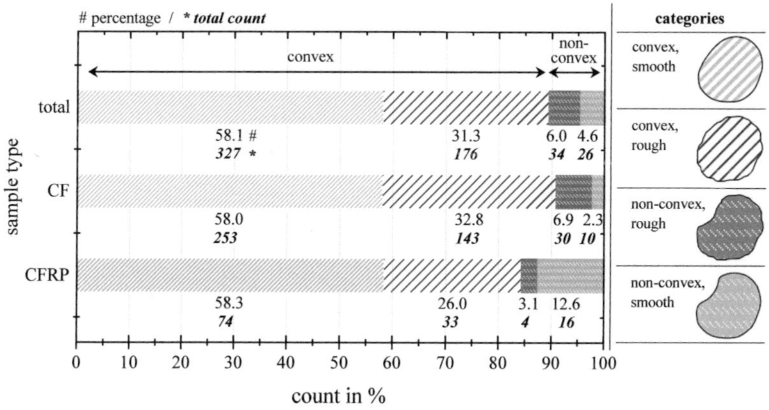

Figure 6.3: Evaluation of a microsection, visually classifying the fibre shapes as either convex or non-convex. In addition, sub-categories smooth and rough are defined. the number given in standard letters refers to the percentage for each category and the total count of fibres is stated in bold and italic. *CF* and *CFRP* are different sample types, *Total* refers to an addition of both samples.

6.1.2 Measurement of strain and the Young's modulus of carbon fibres

The Young's modulus was determined according to the procedure presented in Section 5.2.2. Virgin fibres were tested with gauge lengths of 20, 35 and 50 mm. Recycled fibres could only be measured with a gauge length of 20 mm: the material production described in Section 4.1 resulted in recycled fibres with a maximum length of 50 mm. Due to constraints in handling and clamping, the examinable gauge length was 20 mm for rCF. To interpret the impact of recycling on the fibre properties, vCF and rCF results at a gauge length of 20 mm are compared directly; for vCF possible influences of the fibre length on the Young's modulus are investigated. For comparison, high modulus fibres of type PYROFIL HR40 by GRAFIL INC. and high tenacity fibres type T700 manufactured by TORAY, were tested according to the same procedure. Measurements were run step-wise to allow for high-resolution DIC images to be collected. By DIC, nine strain values were determined and the Young's modulus was calculated for each step. To account for the non-linearity of carbon fibres, each pair of Young's modulus and strain was plotted and a fit function was calculated. The intersection with the Y-axis was defined as resting modulus E_0 and used for comparison. By means of example, this evaluation procedure is displayed in Figure 6.4. It was run for each single fibre.

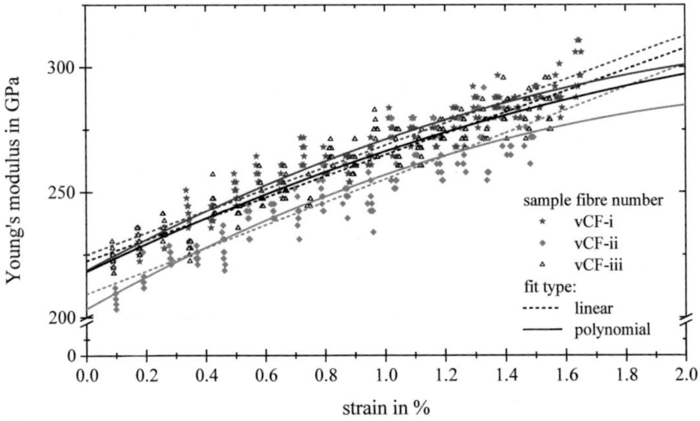

Figure 6.4: Young's modulus as a function of strain regime; exemplary results for four fibres measured with a gauge length of 35 mm tested with the SiFiT. Each family of symbols correlates to one individual sample fibre. Data is fitted by linear (dashed line) and polynomial regression as recommended in [198]. The intersection of the linear fit with the Y-axis is defined as resting modulus E_0 and used for comparison.

In similar plots, Djordjević et al. [198] investigated the non-hookean behaviour of carbon fibres and fitted their strain data by linear and second-order polynomial regression in order to obtain the resting Young's modulus. Both approximations are exemplary shown for three fibres in Figure 6.4. Both fits have been applied to the eleven fibres tested with a gauge length of 35 mm. The apparent Young's modulus obtained by linear regression $E_{0,lin}$ and by second-order polynomial regression $E_{0,2nd}$ as well as the coefficient of determination R^2 are listed in Table 6.1.

Regarding the representation of the Young's modulus in Figure 6.4, one should keep in mind that the large range of variation of the Young's modulus within one cluster is not a consequence of the material properties but stems from uncertainties in the measurement of the strain variation $\Delta\varepsilon$ and then applying Equation 2.1.

Table 6.1: Resting Young's modulus, calculated by linear regression and by second-order polynomial regression. The intersection with the Y-axis is defined as the resting Young's modulus.

linear regression						
fibre	*1*	*2*	*3*	*4*	*5*	*6*
intersection $E_{0,lin}$ in GPa	230.8	214.0	225.0	229.5	209.3	231.2
R^2	0.839	0.876	0.913	0.747	0.869	0.896
fibre	*7*	*8*	*9*	*10*	*11*	
intersection $E_{0,lin}$ in GPa	222.5	218.4	219.5	221.8	223.7	
R^2	0.890	0.896	0.857	0.897	0.831	

second-order polynomial regression						
fibre	*1*	*2*	*3*	*4*	*5*	*6*
intersection $E_{0,2nd}$ in GPa	227.8	210.7	218.8	221.7	203.4	233.3
R^2	0.843	0.880	0.923	0.762	0.879	0.897
fibre	*7*	*8*	*9*	*10*	*11*	
intersection $E_{0,2nd}$ in GPa	218.2	211.9	214.8	219.0	218.4	
R^2	0.896	0.913	0.865	0.900	0.841	

On average, the resting modulus $E_{0,2nd}$ acquired by second-order polynomial regression lies 2.0 % below the resting modulus $E_{0,lin}$ found by linear regression. For each fibre 1 to 11, the coefficient of determination R^2 is up to 2 % higher for the polynomial regression compared to the linear regression. Given this minor difference in R^2 between both procedures, the linear regression was chosen in favour for better comparability [141, 198] and is used for further evaluation in the work at hand. In the following, the resting Young's moduli of individual fibres, determined by linear regression, are plotted versus the cross-sectional area of the respective fibre (Figure 6.5). This illustration is chosen to account for the fact that the Young's modulus and the cross-sectional area correlate inversely [24, 58, 223]. While this correlation is commonly described in terms of fibre population where both quantities are averaged, Figure 6.5 demonstrates that it is likewise found for individual fibres.

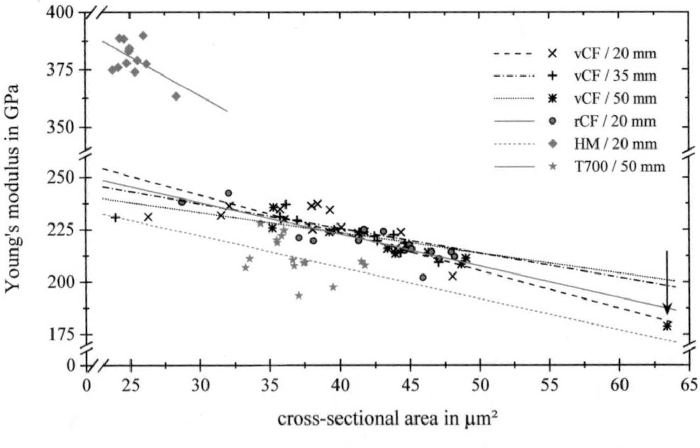

Figure 6.5: Resting Young's modulus as a function of cross-sectional area determined by means of laser diffraction. The resting Young's modulus was determined by interpolation procedure presented in Figure 6.4.

Three clusters of data sets are found in Figure 6.5: one consisting of virgin (*vCF*) and recycled (*rCF*) PANEX 35 fibres, one of fibres of type *T700* and one for the high modulus fibres *HM*. The cluster of PANEX 35 fibres can be subdivided in terms of their origin (virgin or recycled) and gauge lengths. For all six sets, a linear fit calculated to relate the resting Young's modulus with the cross-sectional area is shown. The slopes are almost equal for all sets but HM. Particular attention shall be paid to data point obtained for a very large cross-sectional area, highlighted by the black arrow (↓). This is no outlier, on the contrary, it follows the same relation between E_0 and cross-sectional area. Due to the close proximity of the linear fits of all PANEX fibres, no significant effect of gauge length and origin (virgin or recycled) can be deduced. The dominant factor is the influence of the cross-sectional area: larger cross-sectional areas correlate to lower Young's moduli. Comparing individual fibres of type *T700* and type PANEX 35 for approximately the same cross-sectional, it can be seen that *T700* has a Young's modulus which is about 25 GPa lower. A similar correlation between the cross-sectional area and the Young's modulus is found for both types of fibres.

Within the frame of this work, special attention is paid to changes of the mechanical properties due to the pyrolysis recycling process. To exclude possible influence due to the gauge length, a comparison should be conducted for the same gauge length. When investigating the Young's modulus, a gauge length of 20 mm was the only gauge length available for both, virgin and recycled fibres.

To emphasise that no difference in terms of Young's modulus between vCF and rCF can be found, Figure 6.6 shows an enlargement of the region relevant for vCF and rCF tested at a gauge length of 20 mm.

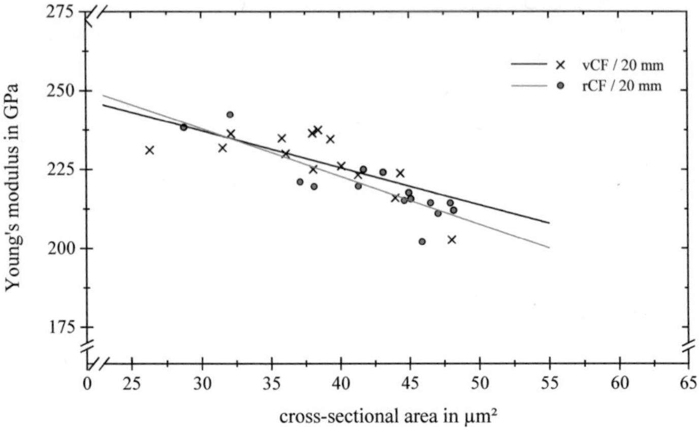

Figure 6.6: Resting Young's modulus as a function of cross-sectional area for recycled and virgin fibres tested at a gauge length of 20 mm.

The point clouds overlap clearly and considering the linear fit functions, no significant difference between population vCF and rCF is found.

6.1.3 Tensile strength of virgin and recycled carbon fibres

As carbon fibres are brittle, the measured tensile strengths exhibit a vast scatter. According to the weakest-link-theory, final failure can be initiated at the most dominant flaw, for instance a surface defect or a large pore. Nonetheless, as pointed out by Bennett et al. [215], it is not necessarily the largest defect that causes failure of carbon fibres.

Hence, tensile strength was evaluated according to Weibull's theory as introduced in Section 2.1.7. The failure probability is plotted versus the logarithm of the tensile strength, exemplary shown in Figure 6.7 for virgin and recycled fibres at a gauge length of 20 mm. The linear fit is then used to calculate the Weibull strength and the Weibull modulus. Tensile strengths showed a unimodal distribution.

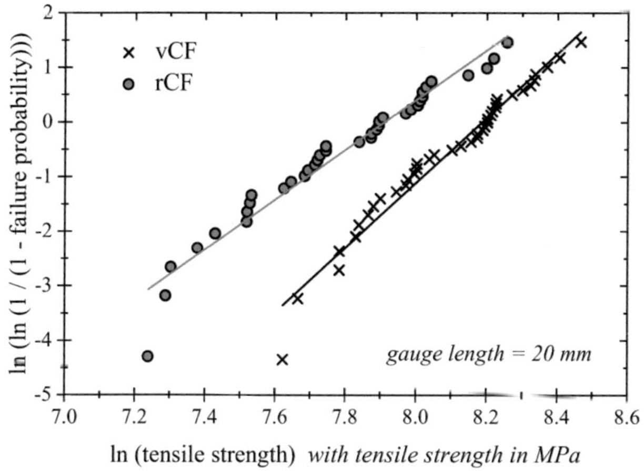

Figure 6.7: Weibull plot of virgin and recycled fibres with a gauge length of 20 mm to determine Weibull strength and Weibull modulus.

The described procedure was followed for different gauge lengths and virgin and recycled fibres. Results are presented in Figure 6.8 and in Table 6.2.

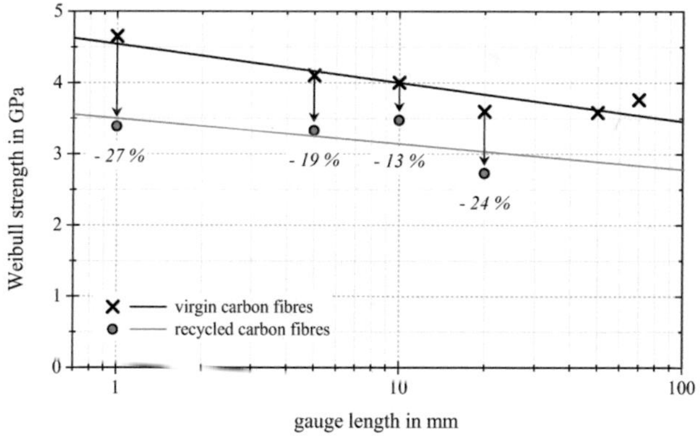

Figure 6.8: Weibull strength of vCF and rCF as a function of gauge length. The numbers in italic display the loss in tensile strength due to recycling. Recycled fibres could not be tested at gauge lengths of 50 nor 70 mm.

For testing virgin fibres, lengths between 1 and 70 mm were covered. The longest recycled fibres reached a length of 50 mm. Due to handling and clamping limitations, the maximum achievable free length was 20 mm.

Table 6.2: Weibull evaluation of vCF and rCF

material	gauge length l *in mm*	valid n	Weibull modulus m	Weibull strength σ_W *in GPa*
vCF	1	36	5.2	4.65
vCF	5	36	6.2	4.10
vCF	10	35	7.3	4.00
vCF	20	41	5.9	3.60
vCF	50	35	6.4	3.58
vCF	70	36	5.1	3.76
rCF	1	60	7.7	3.38
rCF	5	46	5.9	3.33
rCF	10	56	5.1	3.47
rCF	20	37	4.6	2.73

In addition to Weibull strength and gauge lengths, the table list the Weibull modulus m as a measure of variance and the amount of validly tested fibres. Weibull strength is decreased for larger gauge lengths, for both, recycled and virgin fibres. Comparing both fibre origins, an averaged decrease of approximately $21 \pm 6\,\%$ in Weibull strength is found for the recycled fibres. Possible reasons are discussed in Chapter 7.

In summary, recycling had affected the mechanical properties of single carbon fibres: Whilst the Young's modulus of the fibres remained unaffected, the tensile strength of recycled fibres was significantly reduced compared to their virgin equivalent.

6.1.4 Determination of Poisson's ratio or virgin fibres

The Poisson's ratio was determined for virgin fibres. Twelve single fibres were validly tested using the experimental setup described earlier (cf. Section 3.1.4). For evaluation, each cross-sectional area is transferred to an equivalent circular diameters and plotted versus strain. In order to obtain reliable results, a linear fit was used to calculate the reference apparent diameter. The fit is needed as diameter variations between adjacent load steps might fall within the resolution limit of the LDS0200. Three exemplary fit functions are shown in Figure 6.9.

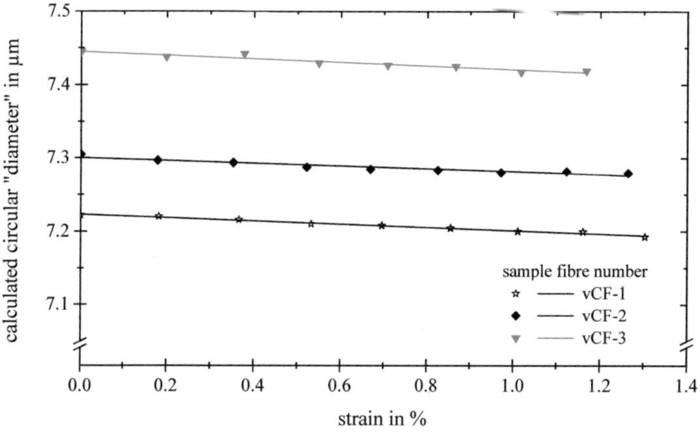

Figure 6.9: Exemplary graphs for determining the Poisson's ratio by plotting calculated diameters as a function of strain. The intersection with the Y-axes represents the reference value of the initial diameter.

The fit equation is:

$$d(\varepsilon_l) = d_{app,0} + \Delta d(\varepsilon_l) = d_{app,0} + a \cdot \varepsilon_l \tag{6.1}$$

with a being the slope of the fit, representing the change of diameter per increment of axial strain ε_l.

Equations 3.6 and 6.1 then yield:

$$\nu = -\frac{a \cdot \varepsilon_l}{d_{app,0}} \cdot \frac{1}{\varepsilon_l} = -\frac{a}{d_{app,0}} \tag{6.2}$$

The results of twelve validly tested PANEX 35 fibres are displayed as box plot in Figure 6.10. The whiskers represent the 1.5 interquartile range (IQR); median and mean value are highlighted by arrows.

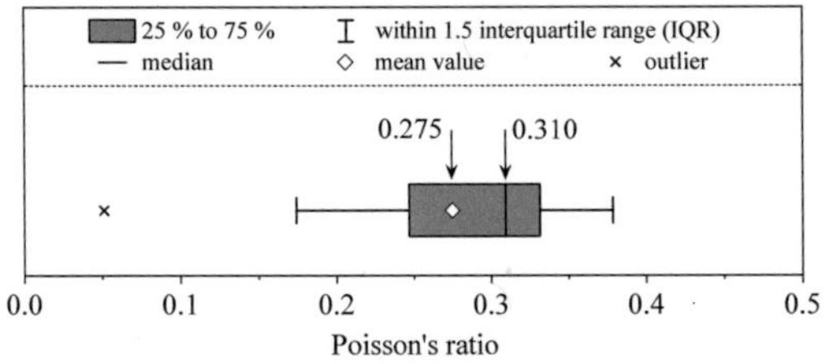

Figure 6.10: Poisson's ratio of PANEX 35, presented as box plot.

6.2 Surface and interface properties

Hereinafter, the term surface properties relates to the appearance of the fibre surface as well as its physical-chemical properties. Results of interface investigations bridge to investigations on composites. In the work at hand, composites were produced with PANEX 35 fibres "as-received". That is, the virgin fibres (vCF) were sized ex works, the recycled fibres (rCF) had lost their sizing during the pyrolysis. Beyond that, the fibres were neither desized nor chemically treated prior to manufacturing and testing. This enables a realistic investigation of the fibre properties and correlations to the composite.

6.2.1 Investigation of the fibre topology by SEM

Results of scanning electron and atomic force microscopy are presented in the following. In addition to the SEM investigations focussing on fracture surfaces of composites (Section 6.3.4), virgin and recycled fibres were investigated to detect possible surface defects or structural changes due to recycling. A qualitative comparison of characteristic features on fibre level is given in Figure 6.11.

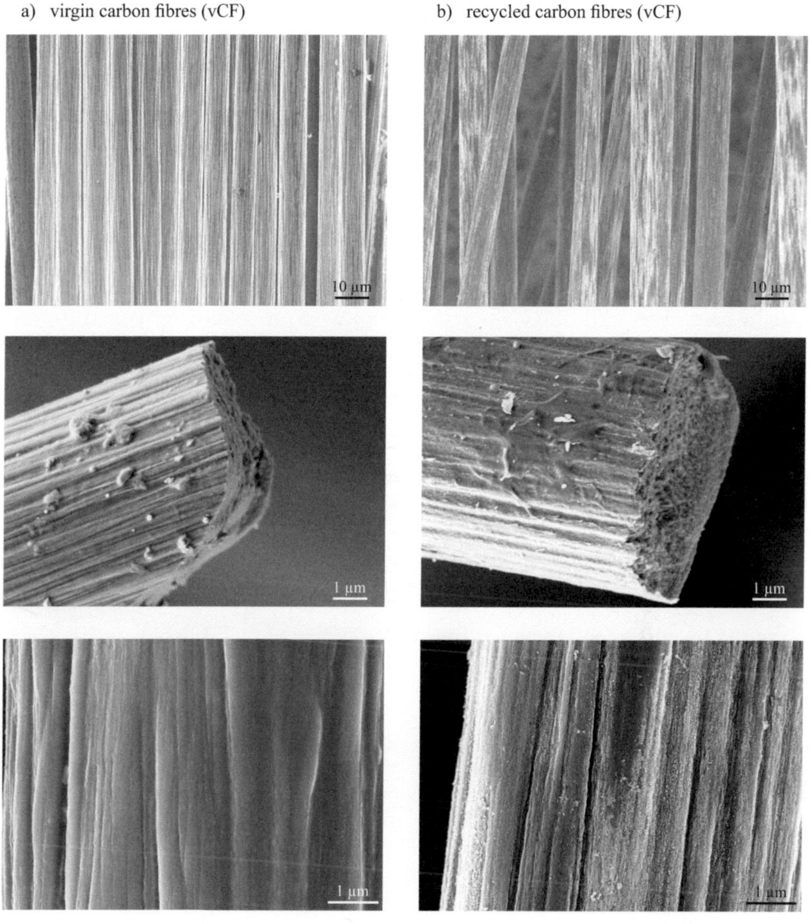

Figure 6.11: Scanning electron micrographs of virgin (left column) and recycled carbon fibres (right column). Magnification and scale are the same for both images in a row. While all images show the crenellations typical for carbon fibres, a nanoscale structure appears for recycled fibres. // *Images 1st and 2nd row taken with SEM/VIE; 3rd row with SEM/LEM.*

The images reveal the typical structure of carbon fibres, showing distinct crenellations along the axis of the fibre. More importantly, they uncover unambiguously damage caused by pyrolysis. The crenellations seem to become more distinct, which becomes particularly clear for large magnifications.

6.2.2 Investigation of the fibre topology by AFM

Atomic force microscopy investigation, run at the at the chair "Experimental Physics II" at the University of Augsburg, were used to investigate the surface of the fibres and to compare virgin and recycled fibres. For both, four fibres were randomly picked and distinctive features are displayed in Figure 6.12. The curvature of the fibre was subtracted to visualise the roughness in relation to base level. All fibres show the lengthwise crenellations typical for carbon fibres [320], also seen in SEM investigations in Figure 6.11. Impurities, presumably dust, are to be found on both fibres. A total of fourteen AFM images from eight individual fibres was considered for quantitative analysis, seven images of virgin and recycled fibres: in each image, six equidistant elevation profiles were placed perpendicular to the fibre direction. For these elevation profiles, maximum roughness R_{max}, roughness depth R_z and mean roughness R_a were determined with the software *Nanoscope Analysis 1.5*. Results obtained for virgin and recycled fibres are displayed in Table 6.3.

Table 6.3: Surface roughness of virgin (vCF) and recycled (rCF) carbon fibres. Results are listed as average value (mean) plus standard deviation (SD).

material	maximum roughness R_{max} in nm		roughness depth R_z in nm		mean roughness R_a in nm	
	mean	SD	mean	SD	mean	SD
vCF	223.7	68.0	106.2	53.9	44.5	15.2
rCF	199.5	49.7	97.7	42.9	39.8	11.8

None of the micro-scale roughness parameters R_{max}, R_z and R_a suggests a significant difference between the roughness of virgin and recycled fibres. However, topographic differences on nanometer-scale are apparent.

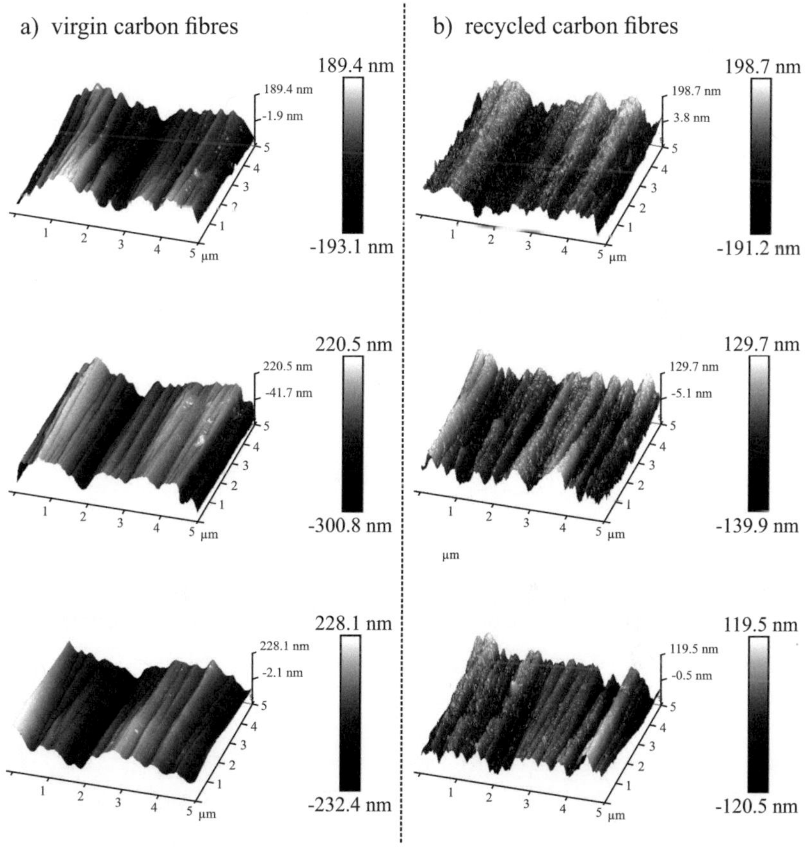

Figure 6.12: Investigation with a atomic force microscope of virgin (a) and recycled (b) carbon fibres. //Images taken at University of Augsburg and visualised in the software *Nanoscope Analysis 1.5.*

6.2.3 Fibre surface properties of virgin and recycled fibres

Physical-chemical properties were determined according to the procedures presented in Section 5.3. In what follows, the results are briefly introduced.

Dynamic vapour sorption shows that recycled state of the carbon fibres is more hydrophilic, as displayed in Figure 6.13. Recycled fibres adsorbed 3 - 4 times more water than virgin fibres.

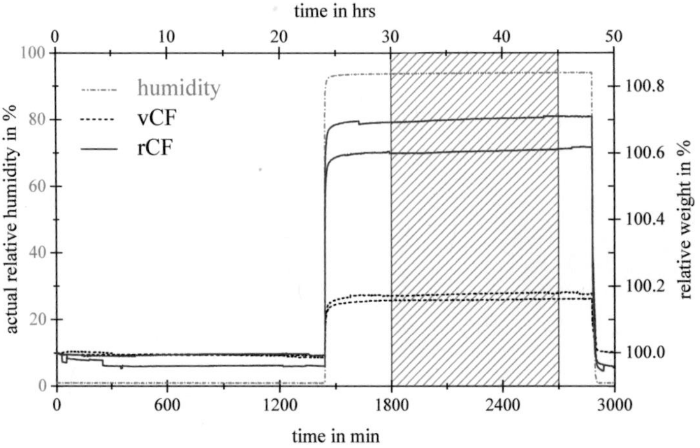

Figure 6.13: Dynamic vapour sorption of virgin carbon fibres and recycled carbon fibres. Values were averaged within the shaded box.

The maximum surface areas of BET measurements were $1.64\,\mathrm{m^2\,g^{-1}}$ for virgin carbon fibres and $3.15\,\mathrm{m^2\,g^{-1}}$ for recycled carbon fibres, respectively.
Results of inverse gas chromatography were evaluated according to the Dorris-Gray method [330] and result for the total surface energy γ_s^T, the dispersive surface energy γ_s^D and the specific surface energy γ_s^{AB} are depicted in Figure 6.14, 6.15 and 6.16, respectively.

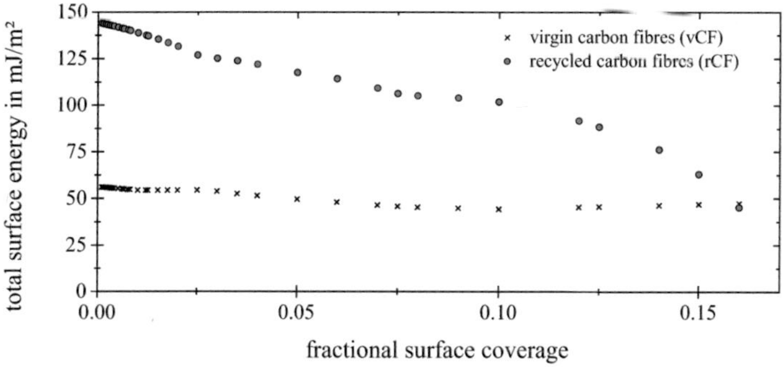

Figure 6.14: Total surface energy γ_s^T as a function of surface coverage.

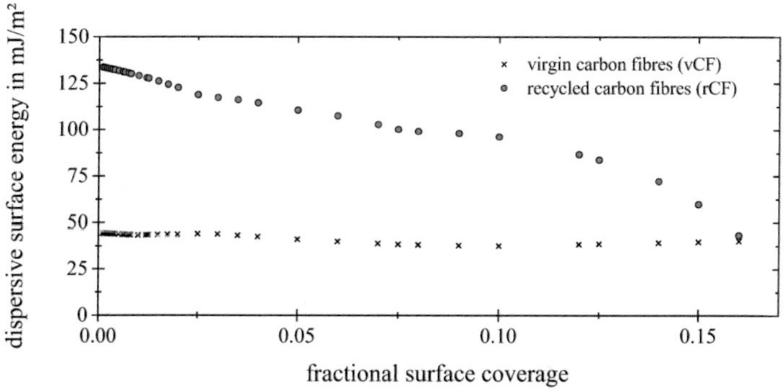

Figure 6.15: Dispersive surface energy γ_s^D as a function of surface coverage.

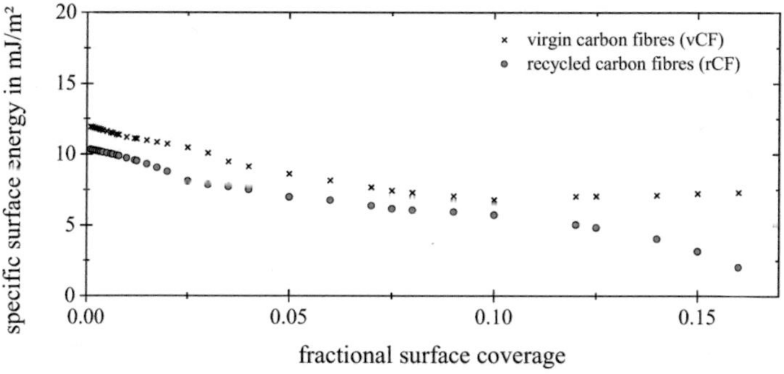

Figure 6.16: Specific surface energy γ_s^{AB} as a function of surface coverage.

The distinct decrease of total and dispersive surface energy as a function of fractional surface coverage imply a heterogeneous surface on recycled carbon fibres.

6.2.4 Fibre/matrix-adhesion: pull-out test

For vCF and rCF, four specimens were tested validly. The data obtained by using the *Fimatest* setup developed by *Textechno* allows to determine the interfacial shear strength τ_{app}, local interfacial shear strength τ_d, interfacial frictional stress τ_f and the critical interfacial energy release rate G_{ic} [322]. Due to the small sample size, all individual data points are displayed in Figure 6.17.

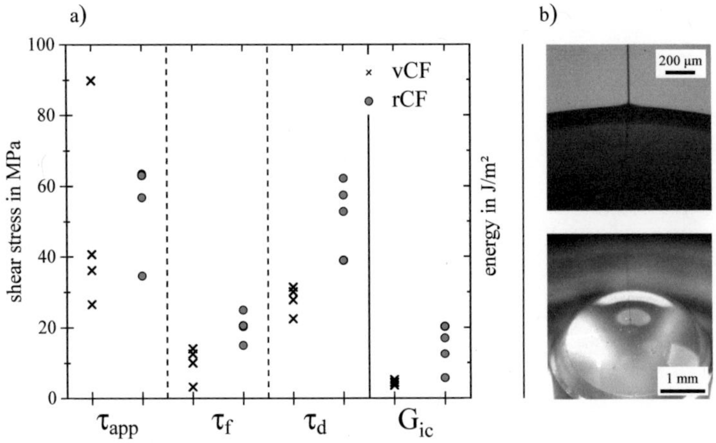

Figure 6.17: Results pull-out test at Faserinstitut Bremen e.V. (FIBRE). Left: comparison of τ_{app}, τ_f, τ_d and G_{ic} for virgin and recycled materials. Right: image of the setup, *provided by Faserinstitut Bremen e.V.*.

The test resin differed from the resins used for composite production within the frame of this work. Still, the following statements can be derived from the pull-out experiment: the interfacial shear strength τ_{app} shows no significant difference between vCF and rCF.

In addition, the following can be stated:

- the interfacial frictional stress τ_f is increased for recycled carbon fibres. Friction between the fibre and the matrix appears to be higher after the maximum of the interfacial shear strength has been reached.
- the local interfacial shear strength τ_d is higher for rCF compared to vCF.
- the critical interfacial energy release rate G_{ic} is increased for rCF, corresponding to an increase in the bonding strength.

6.2.5 Fibre/matrix-adhesion: push-out test

Push-out specimens taken from HS panels were prepared and tested according to the procedures detailed in Section 5.4.2 for samples. Results are displayed in Table 6.4.

Table 6.4: Results of the push-out test with hybrid resin specimens (HS)

material	thickness l_{push} in µm	fibre diameter d_{fibre} in µm	interfacial shear stress τ_{IFSS} in MPa
vHS	22.14	7.81 ± 0.37	72.11 ± 8.43
rHS	34.70	7.62 ± 0.50	67.27 ± 7.53

Given the variation of the results, no significant difference between vHS and rHS in terms of interfacial shear strength is found.

Summarising the results of interface testing, namely push-out and pull-out test, recycling showed no significant impact on the interfacial shear strength. However, in pull-out test an increase in the quantities τ_f, τ_d and G_{ic} was found for recycled fibres combined with a test resin.

6.3 Composite characterisation

This section deals with composite properties, presenting results of quality assessment and mechanical testing.

6.3.1 Process-related material characteristics

Determination of the glass transition temperature by DMA

Three specimens were investigated by dynamic mechanical analysis for each of the four major materials. Two exemplary curves (rHS and rEP) of the loss factor as a function of temperature are shown in Figure 6.18 and results are presented in Table 6.5. Storage and loss modulus are graphed for the sake of completeness but will not be further discussed.

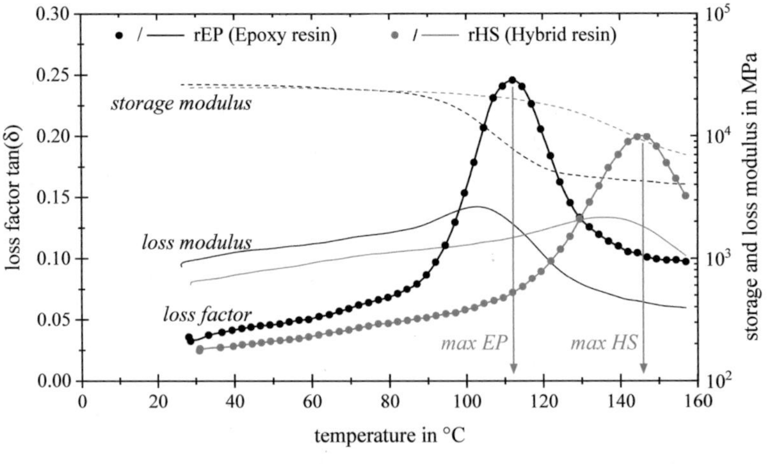

Figure 6.18: Temperature sweep of DMA experiment, exemplary shown for one rHS and one rEP specimen. The maximum of the cubic spline fit of measurements of the loss factor was chosen to determine the glass transition temperature.

Given the low standard deviation of glass transition temperatures, a homogeneous curing of the panels can be assumed.

Table 6.5: Results of the Dynamic mechanical analysis

material	resin	glass transition temperature in °C	standard deviation in °C
vHS	hybrid resin	147.9	0.6
rHS	hybrid resin	145.3	0.4
vEP	epoxy resin	112.7	0.9
rEP	epoxy resin	111.8	0.2

Fibre weight fraction determined by TGA

Thermogravimetric analysis results for vHS, rHS, vEP and rEP are given in Figure 6.19 as mean values and standard deviations.

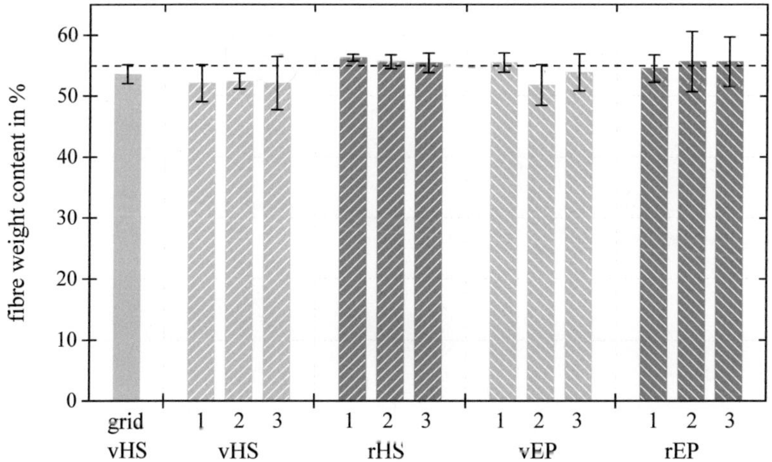

Figure 6.19: Fibre weight contents determined by TGA measurements of panels that have undergone mechanical tests. The number indicate the individual plate from which specimens were cut. Three to six samples were investigated per plate. The bar "grid" refers to the results discussed in Figure 6.20 and averages the result presented there.

The labels 1, 2 and 3 indicate that specimens were cut from three different panels. Panel 1 was used for tensile tests according to "approach 1", panel 2 and 3 for tensile test according to Weibull's theory labelled "approach 2". The label "grid" refers to the averaged result of the 64 measurements displayed in Figure 6.20. The set value of 55 % fibre weight content is shown as dashed line. There is some scatter in the results, nonetheless, all but one error bar overlap with the set value. It was thus deduced, that the fibre weight contents of the panels are sufficiently consistent to allow for comparison of different panels and materials.

Results for the grid of 64 specimens are also displayed as a heatmap in Figure 6.20. In the diagram, Q indicates the position of an overflow channel, which is facing upwards and is being used to ease the demoulding of the panels. The circle labelled with an ∗ shows the scale of the circular 25 mm-specimens.

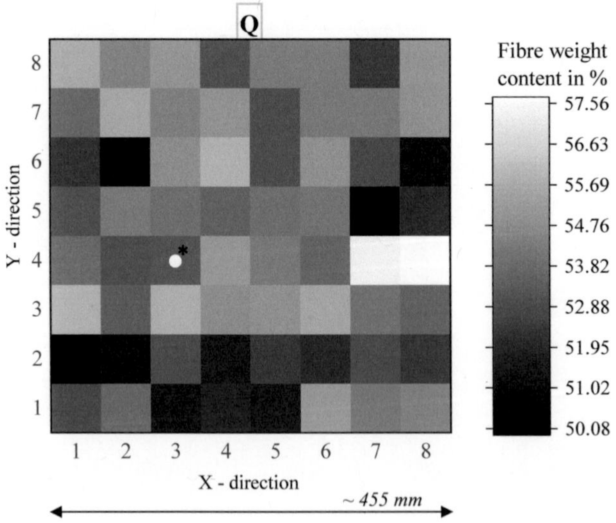

Figure 6.20: Fibre weight content distribution determined by TGA for one panel: The grey-scale squares refer to fibre weight content in wt.-%. The white circle on coordinate [3;4], labelled ∗, represents the size of a circular 25 mm TGA specimen disk. The label Q points out the position of an overflow channel of the panel. In this visualisation, the channel faces upwards and it can be used to demould the panel after production.

The bulk (matured resin-matrix-mix) had been placed in the centre of the mould. The fibre weight content fluctuates but given its distribution, these variations seem to be arbitrary and not a result of the positioning, edge-effects or other physical influences.

Determination of density by gas pycnometry

Results of density measurements by gas pycnometry are presented in Table 6.6.

Table 6.6: Density of fibre and composite materials determined by gas pycnometry

category	fibres			composites			
material	vCF	rCF	CFF	vHS	rHS	vEP	rEP
density in $\mathrm{g\,m^{-3}}$	1.81	1.91	1.87	1.45	1.46	1.45	1.46
SD in $\mathrm{g\,m^{-3}}$	0.01	0.01	0.01	0.03	0.01	0.01	0.02

As can be seen, the mean density of the recycled fibres (rCF) is 5.6 % higher than those of the virgin carbon fibres. The same tendency is found for the composites while no significant difference between both resin systems, HS and EP, is found. For the sake of completeness, ground carbon fibres (CFF) are displayed. This material was used as filler in a holistic recycling approach [251].

6.3.2 Tensile properties of bulk moulding compounds

Tensile properties according to the standard procedure

Tensile properties were investigated according to the standard testing procedure described in the preceding section. The Young's modulus was calculated by linear regression. Due to low strains to failure, the Young's modulus was, deviating from standard DIN EN ISO 527-4, determined between strains of 0.05 and 0.15 %. The tensile strengths and the Young's modulus of all four materials are visualised as bar charts in Figure 6.21. Standard deviation is given as error bars. Validity was defined in reference to Figure 5.2 a) and set as follows:

for evaluating the tensile strength, all specimens which fractured completely within the patterned and the light-grey area were considered; Young's moduli were calculated for all specimens that did not fail in the gripping regions (shaded). Strains were measured by digital image correlation in the patterned area indicated by R in Figure 5.2.

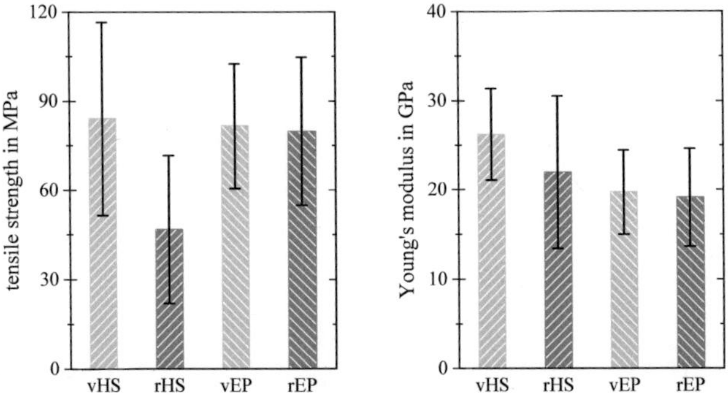

Figure 6.21: Tensile properties according to the standard procedure: tensile strength (left) and the Young's modulus (right); displayed as mean values and standard deviation.

The experimental results shown here exhibit scatter too large to allow for a fair and unambiguous distinction between the different materials. Given the large standard deviation, no significant difference is found between the materials, neither between different resin systems, nor between the virgin and the recycled fibres. Type rHS seems to be reduced in terms of tensile strength but still, the error bars overlap.

Tensile properties according to the Weibull procedure

The tensile tests and additional SEM investigations indicated that failure was initiated in the vicinity of weak spots, namely resin-rich regions and fibre bundles perpendicular to the loading direction. Consequently, considering the failure mode in the sense of a Weibull distribution makes sense. For Weibull and the weakest-link theory, failure will initiate at the weakest spot and not due to geometric features. Therefore, the region-of-interest was extended to the whole clamping length. All specimens where failure occurred between the two clamping regions were defined as being valid, corresponding to the grey area in Figure 5.2b). Conversely, samples were excluded from the evaluation in case failure occurred within the clamps or was initiated by them. Exemplary curves are found in Figure 6.22.

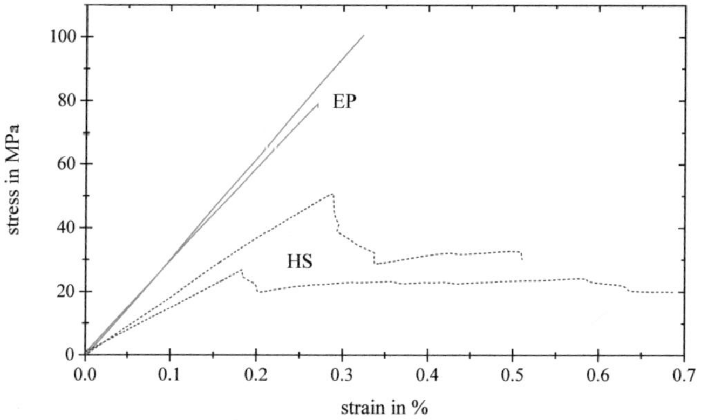

Figure 6.22: Typical stress-strain curves for HS (dashed line) and EP (solid line) materials.

Based on the stress-strain curves and on visual inspections of the failed specimens, two types of failure mechanisms were tendentially detected: The HS specimens tended to fail progressively, with the crack crossing a larger area. In some spots, the crack ramified or delaminations occurred. Very few specimens showed simultaneous failure in different regions. Most of the HS specimens retained their cohesion even after a load drop of 80 % was observed and the experiment had ended. On contrary, EP specimens failed more abruptly and the cracks did not split. The majority of EP specimen failed completely, resulting in two pieces. To quantify the observation made during tensile tests, results for all valid specimens are presented as Weibull plot shown in Figure 6.23; the results and the amount of valid specimens per material are summarised in Table 6.7. Per material, 50 specimens were available and 60 % to 70 % showed valid failure.

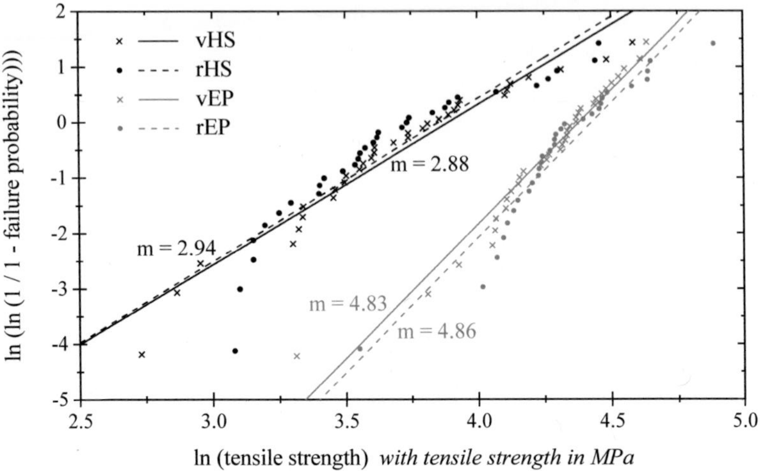

Figure 6.23: Weibull plots for the composite materials.

Two clusters appear: one for resin HS, one for EP. Within each cluster, data points overlap. This observation is underlined by the results in Table 6.7. Weibull strength and Weibull modulus of composites containing virgin or recycled fibres are almost the same but differ for the different resins.

Table 6.7: Results obtained by the "Weibull evaluation procedure" of composites

material	**Weibull strength** in MPa	**Weibull modulus**	**valid specimens**
vHS	48.96	2.88	33
rHS	47.19	2.93	31
vEP	79.83	4.83	34
rEP	83.71	4.86	30

For sake of completeness, the Young's moduli were also determined for the here described Weibull procedure. Due to low strains to failure, the region for evaluating the Young's modulus was set to 0.05 to 0.10 % strain. Results are shown in Figure 6.24, where they are compared to the results obtained by the standard procedure.

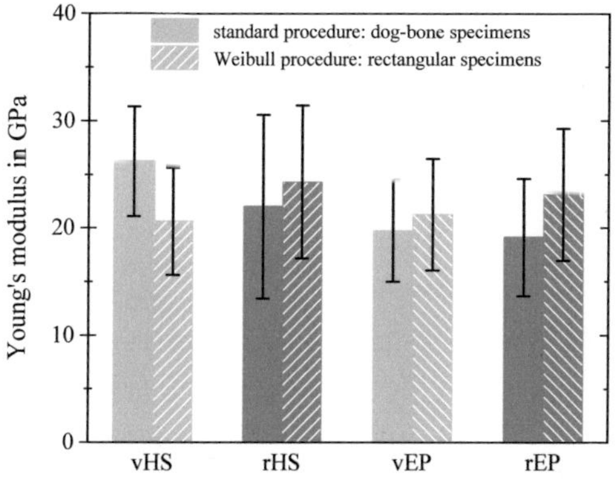

Figure 6.24: Comparison of the Young's modulus of dog-bone and rectangular specimens. No significant effect in Young's modulus is found when comparing the type of specimens and the different materials.

As aforesaid, scatter in the data of the Young's modulus for BMC materials is very high. As all error bars overlap, no significant conclusions can be drawn: neither the resin system, nor the origin of the fibres, nor the testing procedure have an identifiable implication on the Young's modulus.

6.3.3 Analysis of composite fracture surfaces by scanning electron microscopy

In this section, scanning electron micrographs of the four materials are presented. They were chosen as representative selection of a variety of specimens under investigation of each material combination. Figures 6.25 to 6.27 show three different magnifications of the materials vHS, rHS, rEP and vEP (clockwise). All images show fracture surfaces of the composites created by tensile

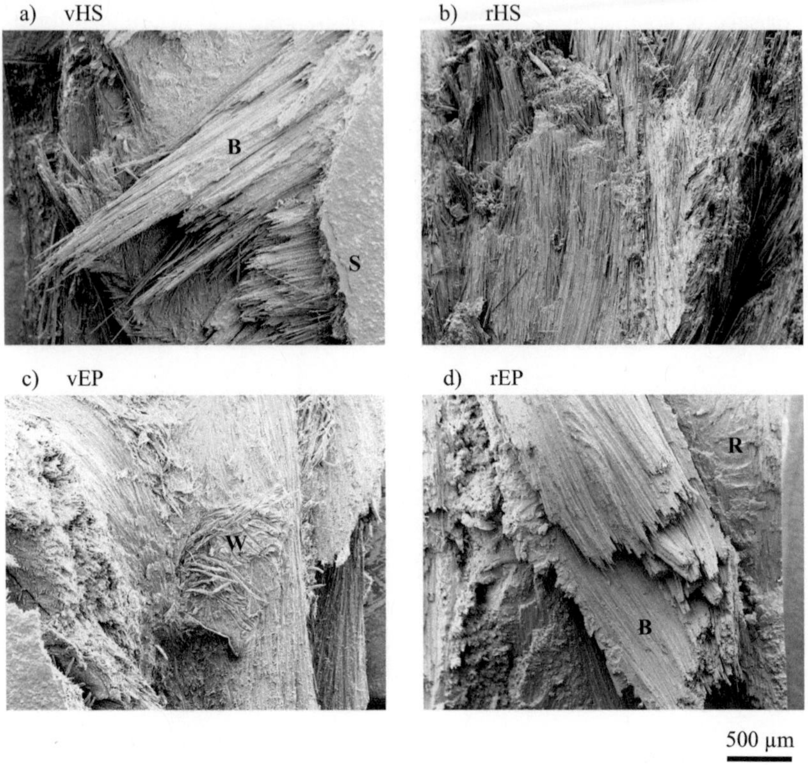

Figure 6.25: Fracture surfaces of BMC composites observed by SEM/IAM - Part I: a) virgin carbon fibres with hybrid resin, b) recycled carbon fibres with hybrid resin, c) virgin carbon fibres with epoxy resin, and d) recycled carbon fibres with epoxy resin. The bundle-like character of BMC specimens can be seen. *B* highlights fibre bundles, *S* the specimen surface, *R* regions rich of resin and *W* a remnant of a stitching yarn used to bind fibre rovings in the original carbon fibre fabric.

tests of dogbone-shaped specimens ("Approach 1", cf. Section 6.3.2). Some features shall be highlighted: all materials exhibit bundle-like characteristics, that is, hundreds of adjacent fibres are running almost in parallel to each other. Cracks run along fibre bundles, frequently to those aligned unfavourably to the tensile direction.

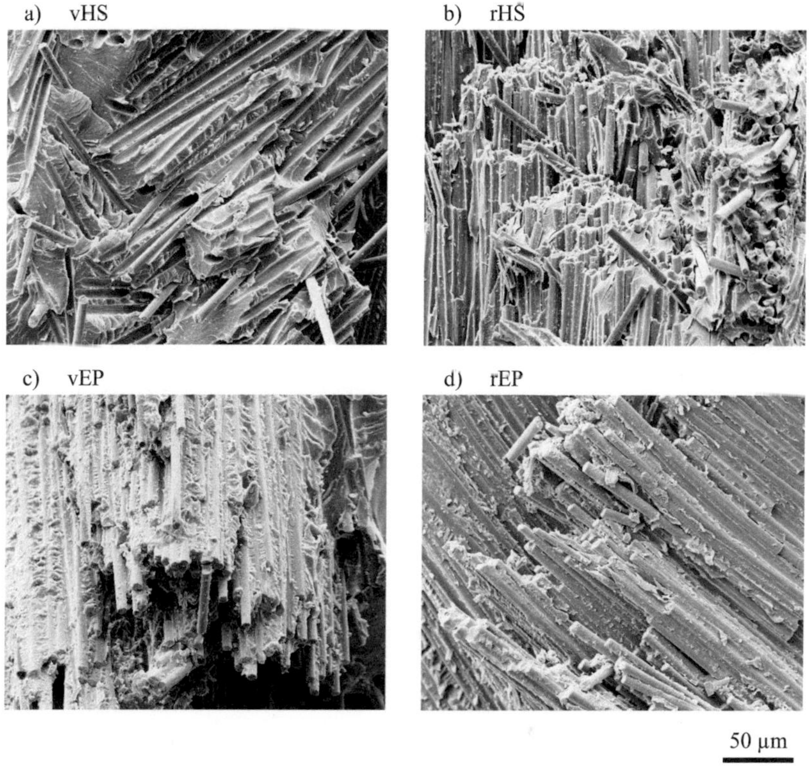

Figure 6.26: Fracture surfaces of BMC composites observed by SEM/IAM - Part II: All images emphasise that fibres appear in bundles in BMC materials. Only few fibres are present as single fibres.

In Figure 6.25, some bundles are highlighted by *B*, whereas resin-rich regions are labelled *R*. In Figure 6.25 a), that is a vHS sample, the flat surface of the specimen *S* can also be seen. Another feature frequently found within the fracture regions of composites containing virgin carbon fibres (vHS and vEP) are remains of the polymer stitching yarns, highlighted by *W* in Figure 6.25 c). The stitching yarns were originally intended to bind the rovings in the fabric patches but are not found in standard BMC material. For vHS and vEP, the

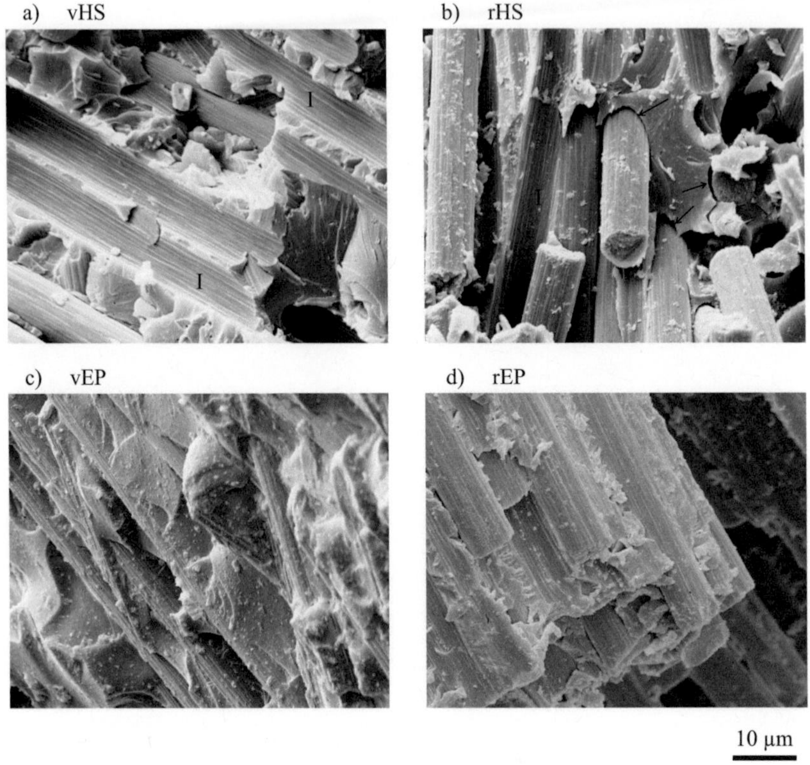

Figure 6.27: Fracture surfaces of BMC composites observed by SEM/IAM - Part III: At higher magnification, the quality of the interface can be assessed. In image rHS, black arrows highlight positions where the matrix detached from the fibre. Commonly, the matrix is marked by imprints *I* in places where fibres lay before fracture. This effect is less pronounced for EP. In addition, more resin residue is found on the fibres in EP composites than in HS composites.

carbon fibre patches were added to the resin without removing the stitching yarns. During the pyrolysis process prior to the production of the composites containing recycled fibres (rHS and rEP), the stitching yarns were destroyed and no residue was found in SEM investigations. In this aspect, the composites based on virgin and recycled fibres differ from each other.

However, the appearance of composites containing virgin and recycled fibres does not differ in terms of their bundle structure. Fibre bundles are also clearly visible at higher magnification, shown in Figure 6.26. Failure seems to occur at the interface between matrix and fibre or fibre bundles, respectively.
Fibre bundles detached during fracture, leaving behind imprints in the matrix as can most prominently be seen in Figure 6.26 a). In both, Figures 6.25 and 6.26, adjacent fibres show almost the same length.
With scanning electron micrographs at even higher magnifications, qualitative observations concerning the fibre/matrix-interface can be made. Frequently, the matrix shows imprints, highlighted by *1*, in places where fibres lay before fracture occurred. Detachments of the fibre from the matrix are also noticeable in image b) (rHS), signalised by the black arrows. Both characteristics were less prone for the EP resin. Furthermore, the fracture surfaces of HS composites (for instance Figure 6.27 a) and b)) contain rather clean fibres and only minor residues. Larger amounts of resin residue seem to be remaining on fibres mixed with EP resin.

6.3.4 Microscopic investigations of BMC specimens

Investigation on microsections of virgin fibre BMC specimens by light microscopy, shown in Figure 6.28, reveal several aspects: as already highlighted by Figures 6.25 to 6.27, region rich in matrix (R) and stitching yarn (W) are found in the BMC materials.
Furthermore, bundles are identifiable and numbered in Figure 6.28. Several fibres are in direct contact but spaces among them are thoroughly filled with matrix. All bundles appear to be impregnated well. While few single fibres are found to be separated from the original fibre bundles, the majority of fibres is still clustered in bundles in the cured composite.

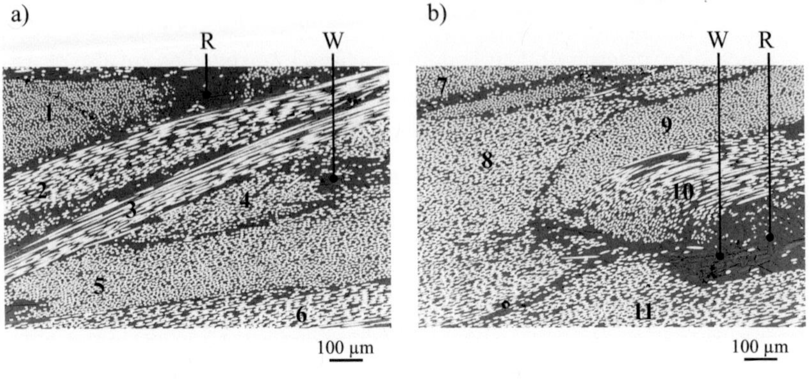

Figure 6.28: a) and b) Microsection of virgin fibre BMC composite: individual roving are distinguishable (1 to 11). In addition to the bundle-like clusters, polyester stitching yarn (W) and regions rich in resin (R) are found.

7 Discussion: Recycling induced properties of carbon fibres and their composites

In the preceding chapters, a method for advanced single fibre testing has been presented and its applicability has been thoroughly discussed (Chapter 3). The tested fibre and composite materials have been introduced (4), experimental procedures have been reported (5) and the results have been presented (6).
In this chapter, the effect of recycling on fibre and composite properties, here BMC, is discussed. Where needed, arguments concerning recycling are accompanied by statements regarding the reliability of the experimental procedures.

7.1 Discussion of single fibre properties

To compare the impact of recycling on fibre level, it was decided in favour of single fibre tests rather than fibre bundle test as the latter exhibits disadvantages: for instance, the shock caused by the first fibre failure can initiate further breakages in adjacent fibres [95], leading to a progressive failure mechanism [96]. Comparing single fibre and bundle testing on cotton, it was found that the single fibre strength is the upper limit while the bundle strength is lower [97–99]. Friction between fibres is another downside of bundle testing [101]. These circumstances might superimpose effects caused by recycling. Furthermore, the recycled fibres were only loosely connected to each other, so that testing a roving of virgin carbon fibres would have differed too much from testing a recycled bundle. Single fibre properties are discussed in what follows. The focus lies upon the geometrical shapes and the mechanical properties of carbon fibres.

7.1.1 Cross-sectional areas of carbon fibres

Recycling had no significant impact on the cross-sectional area of carbon fibres (cf. Figure 6.2). Thus, this section is decoupled from the discussion of recycling effects. It focusses on the question what procedure to use when investigating the cross-sectional areas of carbon fibres prior to mechanical testing.
A comprehensive literature review of carbon fibre shapes was presented in Section 2.1.5 and summarised in Table 2.2. It was derived that only some types of carbon fibres are circular. By contrast, many fibres have different shapes, as they can for instance be hollow, kidney-shaped or oval. Perfectly circular fibres exist [136, 137] but cannot be taken for granted.
Non-uniformity was also found in the fibres investigated in the work at hand, namely high tenacity carbon fibres of type PANEX 35. It was found experimentally that the cross-sectional areas of PANEX 35 vary greatly from fibre to fibre (cf. Figure 6.3). In general, their fibre shape can be described as oval and approximately 90 % of the samples investigated fulfilled the definition of being convex. A minority of fibres exhibited slightly kidney-shaped characteristics. In this context, the importance of determining individual cross-sectional areas instead of averaged diameters was emphasised.
The developed testing procedure to determine cross-sectional areas relied on results obtained by using a laser diffraction sensor. The technique was originally invented to gauge diameters and eccentricity of wires [160–162] and is nowadays used for investigating fibre properties [163]. Instead of designing an in-house sensor as reported in [69, 171–173], the commercially available LDS0200 by CERSA-MCI was in use as in several other studies focussing on carbon fibres [179–183].

This is beneficial as the proper functionality of the sensor has been verified before. However, the initial calibration relies on a factory setting based on thin steel wires. Within Section 3.2.2, a comprehensive experimental study concerned with the calibration of laser diffraction sensor for carbon fibres has been discussed. Literature in this field reports deviations between laser diffraction measurements and comparable techniques, such as SEM. While [173] found no systematic error, in some publications, laser diffraction overestimated [176–178], in others, it underestimated [172] the results for fibre diameters obtained by alternative procedures, for instance SEM. This underlines the importance of specifically calibrating laser diffraction sensors to the material under investigation. It was found that a correct calibration is crucial to avoid measurement errors. Nonetheless, many publications neglect this fact and do not provide information of the calibration procedure. For instance, laser diffraction was used for investigations on carbon fibres [110, 179–183] and for natural fibres [187, 188], without providing information of the calibration of the diffraction sensor. Stating more details of the calibration procedure in scientific publications would promote their replicability. In the work at hand, all values were corrected by Equation 3.9 to take into account that the original calibration of the LDS0200 was performed on metal wires. Beyond, publications do often not provide details of the calculation of fibre diameters.

An alternative to laser diffractometry would have been vibroscopic measurements, for instance with the help of the semi-automatic tensile tester FAVIMAT+ produced by *Textechno* [107]. The device has been used to study technical fibres [106, 193] and to investigate recycling effects [108, 196]. The legitimate concerns raised by Titze and Hunter [197] about applying vibroscopy for non-uniform strings can be transferred to carbon fibres. It is questionable whether the vibroscopic principle can be transferred to carbon fibres without adjusting the equation [192] used in the theory of vibration of perfectly flexible strings. According to a short review by the manufacturer of the FAVIMAT+, *Textechno*, it can be used [107]. Nevertheless, no prove of the applicability

is given and the measured linear density is directly linked to a cross-sectional area without taken non-circular fibre shapes into account [107]. Due to these uncertainties, vibroscopic methods were not considered in the work at hand.
Analogously, optical methods such as optical and scanning electron microscopy were not considered for in-situ measurements during tensile tests. They were exclusively applied for qualitative purposes and for comparison as recommended in the standard [1]. In this aspect, laser diffraction sensors gain an edge over scanning electron microscopy, as the former can readily be implemented in mechanical single fibre testing procedures. One should also bear in mind that a superimposed roughness (cf. Figure 6.12) would be beyond the resolution limit of the optical microscopy [320].
Reviews established an explicit correlation between tensile properties of carbon fibres and their cross-sectional areas, whereby reduced cross-sectional areas commonly correlate with an increase in tensile strengths and the Young's moduli [24, 58, 223]. Considering this, enhanced measurements of the cross-sectional area are worth striving for when investigating the structure-property relationship of carbon fibres. Whilst SEM measurements were found to be in the same range as measurements by LDS (cf. Figure 6.2), the latter are expedient to determine the cross-sectional area of fibres prior to mechanical tests.

A computer-assisted approach to compare circular and elliptical approximations was presented in Section 3.2.2. This was of interest, as a true value could be defined. In the following, the elliptical approach is compared to circular assumptions on base of data gathered by means of laser diffraction. Here, the results cannot be compared to a true value but the relative difference can be investigated.
As described earlier, the fibre and the diffraction sensor are not rotated in the *Dia-Stron* testing device LEX820/LDS0200. Instead, one measurement of the apparent diameter is taken and the cross-sectional area is assumed to be circular. In the work at hand, a procedure to approximate an elliptical fibre shape was introduced. Hence it is worth comparing both procedures. As data

base, 200 randomly picked virgin fibres used for tensile testing were chosen. Their cross-sectional areas were calculated according to the developed elliptical approximation procedure and defined as true or nominal values. For comparison, the readings of apparent diameter # 1 (0°-direction) and apparent diameter # 5 (40°-direction) were used to calculate two different circular cross-sectional areas. As the fibres had been transferred to the machine without picking any preferential direction, the values at any arbitrary angle can be assumed to be random. Thus, they correspond to any reading given by the *Dia-Stron* testing device. The relative deviation of the circular assumption was calculated and is plotted in Figure 7.1.

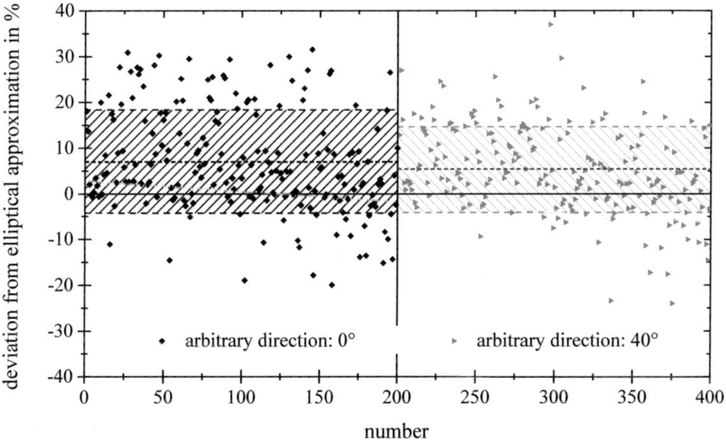

Figure 7.1: Measurement of cross-sectional areas: deviations of common circular approaches from the developed elliptical approach. Values 1 - 200 and 201 - 400 depict two arbitrary directions.

A visualisation as point clouds including the mean value of the relative deviation (dashed line) and the standard deviation (shaded box) was chosen to emphasise several aspects. Both clouds are quite similar, underlining the effect that the angel was randomly chosen. The mean value of the circular assumption lies 5 to 7 % above the elliptical approach. This meets the expectation, that an area calculation based on a single measurement will overestimate the area and underestimate the tensile strength. Apparent diameters, no matter if determined by laser diffraction or side views in SEM, tend to overestimate the real diameter, for instance due to undercuts or unfavourable orientations. Thus, individual values can lie below the real cross-sectional areas, but statistically, more apparent diameter readings will overestimate the reality. It needs to be emphasised that the scatter is very high and circular calculations vary up to 40 % compared to the elliptical approach. According to the reliability study presented in Section 3.2.2, deviations from the actual value of up to 50 % are to be expected. The large deviations raise doubts about the reliability of the circular approach, which is commonly applied in scientific investigations [179–182]. The reliability of the here developed elliptical approach and its comparison to circular assumptions was discussed in detail in Section 3.2.2: there it was concluded, that the elliptical approach is a valuable alternative to state-of-the-art circular approaches.

Based on this comparison, several aspects shall be emphasised here: it is advisable to determine more than one apparent diameter per individual fibre, rotating the laser diffraction sensor with respect to the fibre axis [143]. The procedure used, particularly the axial position, the angular position and the amount of individual measurements should stated in addition to the mere diameter results: rotational increments of 10° [110] and 36° [143] were reported, 15° were recommended in the standard [1]. It was stressed that the last measurement should be 180° minus the rotational increment to avoid multiple readings [1, 143].

It is also expedient to study the fibres before applying laser diffraction: several bundles or fibres can be embedded in potting resin to be ground and polished perpendicular to the fibre direction, according to the standard [1], method C. An analysis of the polished faces should be run to evaluate the most common fibre shapes. In case they are circular, apparent diameters can be used to calculate the cross-sectional area of a circle. Oval fibre shapes should be approximated by using several diameter readings and an elliptical fit [110].

It is not recommended to apply laser diffraction in case the fibre shapes are complex, hollow or kidney-shaped (cf. Table 3.4).

Altogether, it could be demonstrated the elliptical approximation procedure developed in the work at hand gains an edge over state-of-the-art circular assumptions. On this basis, cross-sectional areas were determined via laser diffraction and an elliptical approximation procedure for each individual fibre prior to mechanical testing.
Variations of the cross-sectional area along the fibre axes were investigated and presented in Figure 3.15. As they were sufficiently low, it was deduced that a measurement of the cross-sectional area in the centre of the gauge length is valid. The experimental setup designed within this work, the so-called SiFiT, did not allow for finding the fibre section with minimal cross-sectional area. However, this fact is also not considered in any other work applying state-of-the-art test equipment like the aforementioned LEX820/LDS0200.

7.1.2 Mechanical properties of single carbon fibres

Young's modulus and non-Hookean behaviour of single carbon fibres

As presented in Section 3.1.4, a step-wise procedure was executed to measure strains and calculate the Young's modulus of single carbon fibres. Experimental results showed that deviations between set stress levels and the averaged experimental stress values on each plateau were negligibly low. During the step-wise experiments, possible strain-rate dependencies when increasing the load levels were assumed to be negligible. For instance, Zhou et al. [314] found that the population of HT-carbon fibres type T700 is insensitive to different strain rates in fibre bundle tensile tests.
The step-wise procedure is ideal to investigate the non-Hookean stress-strain behaviour of carbon fibres, first published in 1968 [2]. The Young's modulus has been calculated using a linear and a second-order polynomial fits and the

results were presented in Table 6.1. Given the marginal difference between both R^2-values and considering the overlapping range of variation, the linear procedure was chosen in favour for better comparability [141, 198]. It was used for further evaluation in the work at hand. Applying the linear fit, the modulus was on average increased by 4.9 ± 0.9 GPa per 0.1 % increase in strain. With slopes of 4.9 and 5.4 per 0.1 % strain, similar results of the non-Hookean effect for high tenacity carbon fibres were reported in literature [198].
The impact of composite recycling via pyrolysis on the Young's modulus of carbon fibres was investigated with a set free fibre length of 20 mm to exclude possible effects due to variations in gauge length. Furthermore, no other free fibre lengths were feasible for rCF: as the production/recycling process involved cutting the fibres to lengths of 50 mm and as 30 mm were needed for handling, longer gauge lengths could not be investigated. Shorter gauge lengths were of no interest as an overlap of the bead regions, needed for digital image correlation, would occur. Both aspects are visualised in Figure 3.3.
To allow for a fair comparison between both origins of fibres, the influence of the cross-sectional area of the fibre must be taken into consideration. The Young's modulus varies as a function of mean diameter of a fibre population as shown in Figure 2.9 and described by several authors, for instance in [24, 58, 146, 223, 224]. Moreover, when discussing the mechanical properties obtained by single fibre tensile testing, the cross-sectional area of each individual fibre needs to be considered. A plot of the resting Young's modulus E_0 vs. the cross-sectional area determined by laser diffraction is plotted in Figures 6.5 and 6.6.
With regard to Figure 6.6, one might suggest that vCF has higher moduli than rCF. This assumption simply results from the random selection of samples and the fact that coincidentally, the rCF considered for testing were slightly larger. The minor variation in Young's modulus comparing vCF and rCF relates to the dependency of cross-sectional area and modulus [24, 58, 223].
An impact of recycling on the Young's modulus was not expected: in [64], it was found that the Young's modulus was only marginally affected by the pre-

cursor spinning conditions, concluding that the Young's modulus is an intrinsic property of carbon fibres. From this it can be deduced that recycling should also not severely affect the Young's modulus as process temperatures during pyrolysis are far below the temperatures during production [57, 58]. Thus, the internal microstructure is not influenced and a decrease of the Young's modulus was not expected. In contrast, the fibre surface is manipulated, resulting in the reduction of tensile strength (Figure 6.8) [64].
Several publications were summarised by Pimenta and Pinho [10]: reductions in the Young's modulus of recycled fibres by up to 13 % were found. However, strains were not measured directly, rendering comparison with those obtained in the work at hand inconsistent.
Again, not only recycling effects on carbon fibres are of interest here. In addition, literature shall be reviewed here to discuss the method development in single fibre testing. To allow for a comparison, cross-sectional areas have to be transferred to diameters assuming a circular fibre shape. Vice versa, diameters as stated in literature can be calculated as circular cross-sectional areas. Averaged diameters derived from Figure 6.2 are listed in Table 7.1.

Table 7.1: Transformation or cross-sectional area to diameter by assuming a circular fibre shape

	LDS		SEM	
	vCF	rCF	vCF	rCF
cross-sectional area in µm^2	39.58	38.98	42.72	40.85
diameter in µm	7.10	7.04	7.38	7.21

Given the large variation in diameter (cf. Figure 6.2), the values are in the range literature suggests for virgin PANEX 35, where 7.2 µm and 7.3 µm are reported in the datasheet and in [106], respectively.

In Figure 7.2, the values collected from data sheets are compared to experimental data for the high tenacity fibres under investigation, PANEX 35 and T700S. All experimental values are resting moduli. For the data from literature, the strain regime in which the modulus was determined, is unknown. Presumably, it was not a resting modulus.

As carbon fibres show non-Hookean strain behaviour [2], values given in literature can only be compared correctly if the resting modulus is considered or if the same strain interval is investigated. If the non-Hookean behaviour is not considered, calculations of Young's modulus are overestimating the actual (resting) Young's modulus, as can be concluded from Figure 2.8. Any arbitrary Young's modulus E_ε determined at a strain ε is higher than the resting Young's modulus [141, 200].

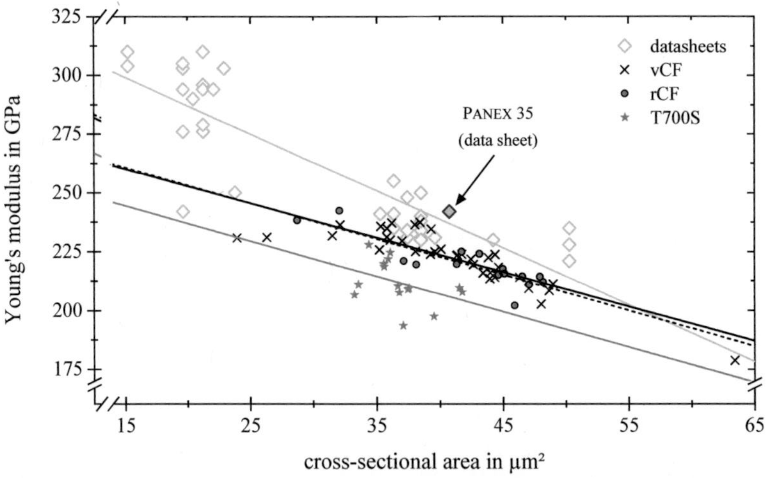

Figure 7.2: Young's modulus vs. cross-sectional area, comparison of data sheets and experimental results (vCF, rCF). The afore described correction function for cross-sectional areas (Equation 3.9) has been applied to obtain the experimental values. The same data sheets as in Figures 2.9 to 2.12 are displayed here.

Highlighted by the arrow is the value for carbon fibre PANEX 35, as found in the datasheet. A strong dependency between the cross-sectional area of carbon fibres and their Young's modulus was observed in experiments and presented in Figures 6.5, 6.6 and 7.2. However, in literature this fact is solely discussed globally for certain fibre populations but not for individual fibres [24, 58, 223]. The same is true for datasheets of manufacturers. Datasheets only contain averaged values of the Young's modulus and the diameter but no correlation between both quantities of individual fibres. The significantly higher slope of the trend line conjoining values from the data sheets is due to the fact that it includes HT and IM fibres, whereas the experimental data consists solely of HT fibres.

Unlike the experimental data, the value for PANEX 35 stated in literature is an average value and lacks consideration of the cross-sectional area and the strain region. A comparative representation is given in Figure 7.3.
Disregarding variations of the cross-sectional area is acceptable in case diameter variations are low. A review by Liu and Kumar [331] stated that within a tow, individual fibres vary by less than 5 %, typically even less than 2 %, between individual fibres. Nevertheless, experimental results in the work at hand show higher variations of the cross-sectional areas, for instance shown in Figure 6.1 b) and Figure 6.2. This might be due to the collection of fibre samples from different tows but the results also suggest that PANEX 35 fibres show a large variation in terms of cross-sectional areas.
To allow for a direct comparison, the results introduced in Figure 6.5 were processed as follows: cross-sectional areas were compared to apparent diameters based on assuming circular fibre shapes. Using the linear fits, Young's moduli for specific diameters, namely 7.0, 7.1 and 7.2 µm, were excerpted. These values are graphed together with data from literature in Figure 7.3.

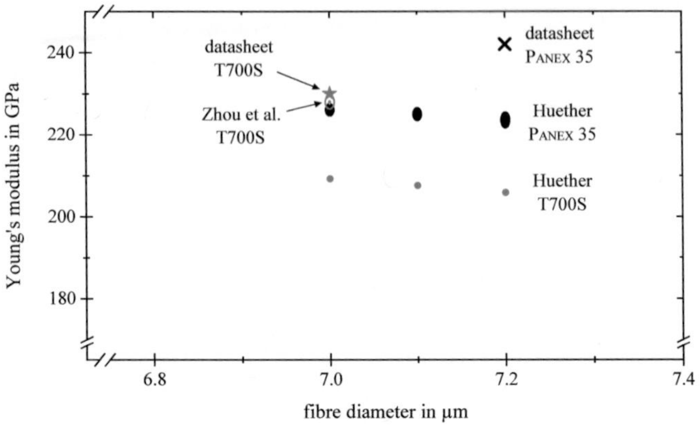

Figure 7.3: Young's moduli of carbon fibres compared to literature as a function of apparent diameter: values obtained experimentally (solid ovals) compared to literature (open oval) [314] and datasheets (grey star, black ×).

Results obtained within this work are labelled "Huether". For PANEX 35, the black ovals span values obtained for three different gauge lengths between 20 and 50 mm; for T700S, solely one gauge length, 50 mm, was available and results are plotted as grey circles.

Only few comparable values are available: a range found in [314] obtained by bundle tests, is displayed as open grey symbol. Datasheets are included as grey star (T700S) and black × (PANEX 35). T700S were also investigated in single fibre tests, and a Young's modulus of 229 ± 12 GPa was reported [332]. Fibre diameters were determined according to standard BS ISO-11567 in SEM [1]. However, instead directly correlating the mechanical properties with the individual diameters, both values were averaged. Thus, no correlation was found and the results cannot be graphed in Figure 7.3. Considering the results of the work at hand, it can be assumed that the large standard deviation is a consequence of the missing correlation with the fibre diameter. Interestingly, the study by Jiang et al. [332] was also concerned with recycling: fibre reclamation by means of supercritical n-propanol did neither have a significant impact on the modulus, nor the Weibull strength.

This comparison should be treated with caution as the data sheets remain short on the question which protocols were used to determine the Young's modulus and the fibre diameter; for instance, no information concerning gauge length and test standard was provided.

It is of interest whether the correlation between Young's modulus and cross-sectional area can be derived from the microstructure of carbon fibres. The presence of a skin-core effect with a denser crystallite packing in the skin is a well-established notion [32–34]. The skin-core concept is supported by state-of-the-art TEM investigations [35–37]. The following concept can be deduced to correlate the skin-core effect and the structure of carbon fibres to their Young's modulus. It can be assumed that highly aligned, densely packed graphite crystallites with low interlayer spacings will contribute more to the tensile modulus than misaligned, amorphous or porous regions.
For estimation, one could assume that a carbon fibre itself is a composite, consisting of highly aligned crystallites (C) and amorphous region (A). In this model, the Young's modulus of C is set to 1060 GPa, the maximum value reported for graphite [17, 18]; and the Young's modulus of A is arbitrarily set to 1 GPa, by assigning typical polymer matrix properties to it. If a load was acting on this idealised "composite" in the direction of the preferred crystallite alignment, a parallel connection can be calculated. In each segment of the fibre ("composite"), the area share A_C attributed to the highly ordered regions:

$$A_C = \frac{E_{mes} - 1}{E_{gra} - 1} \tag{7.1}$$

can be calculated, with E_{mes} and E_{gra} being the Young's moduli experimentally measured and that assumed for graphite, respectively. Applying Equation 7.1 to the results obtained experimentally for virgin carbon fibres, only a share of 0.21 ± 0.01 % is responsible for the measured modulus E_{mes} of carbon fibres. This estimate is clearly too simple to model a real carbon fibre, but can be used as appraisal that carbon fibres show potential for much higher moduli.

Interestingly, Guigon et al [21] estimated that a share of 20 % of the total area of a cross-section of their investigated carbon fibres accounted for the skin. It needs to be mentioned, that other groups found different extents of the skin region [35, 37, 48]. Notwithstanding the fact that such a simplified approach cannot model real carbon fibres, it is deduced that an increased alignment of the BSU accompanied by a decrease in cross-sectional area would result in an increase in the modulus of carbon fibres [33, 333].

A random carbon fibre sample was investigated by transmission electron microscopy to reconstruct whether a skin-core structure existed and if it the skin accounted for an areal share in the magnitude of 20 %. The sample was a virgin fibre and is displayed in Figure 7.4. The upper row shows the oval shape of the carbon fibre (left) and a zoom to mark the regions-of-interest (right). ROI 1 is right at the surface of the carbon fibre, ROI 2 approximately 100 nm away from it. Moving inwards, ROI 3 has a distance of approximately 800 nm, and ROI 4 of 1500 nm to the fibre surface, respectively.
Close to the surface, voids are clearly seen in TEM images. They are up to 20 nm diameter in the plane and seem to be elongated in the fibre axis. The voids are adjacent to regions of highly-ordered crystallites as shown in [334]. High orientation in proximity of voids was also described by Bennett [38]. Apart from that, all regions are more or less akin. Partly, highly orientated graphite layers are present, exemplary highlighted by white boxes. Regions of high orientation are surrounded by regions of rather amorphous carbon. Referring to this random sample, no significant differences between skin and core regions are discovered, or, the skin is thicker than the 1500 nm-layer that could be investigated. For instance, [35, 46] found the skin layer to exceed a thickness of 2 µm. The information revealed in the TEM images within the work at hand is not distinct enough to draw detailed conclusions about crystallite size and orientation, as for instance described in [35]. It must be concluded that the TEM investigation does not support the estimation stated above.

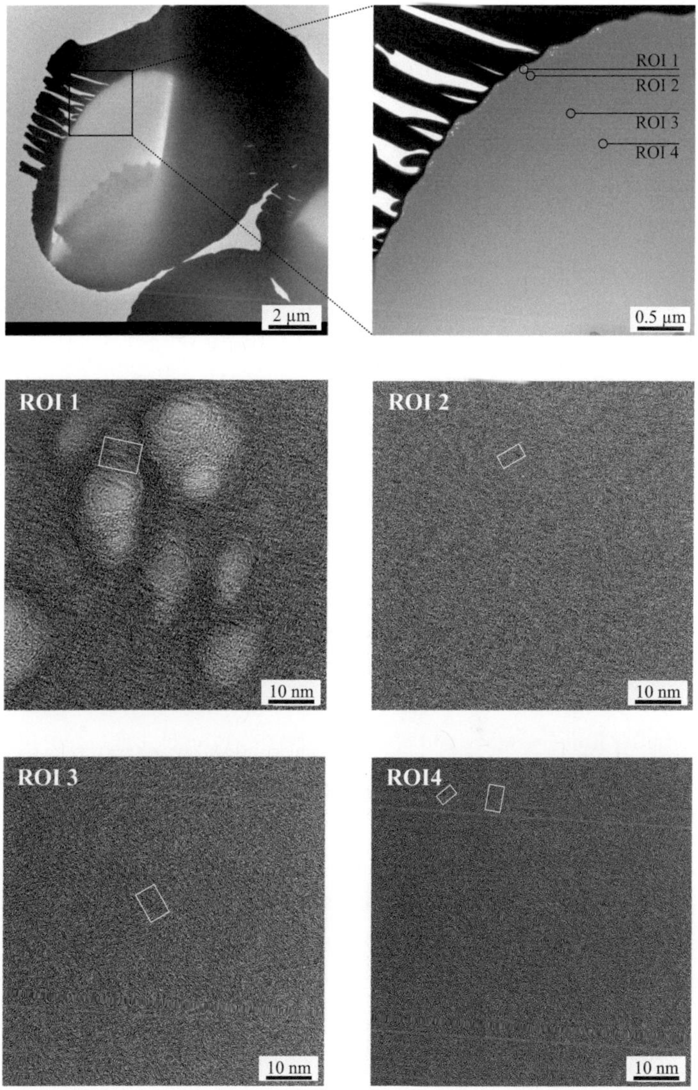

Figure 7.4: TEM images of a virgin carbon fibre, type PANEX 35. Images taken in different regions of interest (ROI), moving inwards. Pores are found in the surface-near region (ROI 1). Highly ordered regions are highlighted by white boxes; however, no distinct skin-core effect is evident.

One attempt to obtain high-performance carbon fibres was made by Morris et al. [46] using very high molecular weight PAN as carbon fibre precursor. It was reported that based on this precursor, carbon fibres with diameters of 2.2 to 2.8 μm and Young's moduli of 361 $\pm$ 45 GPa were produced. It is worth integrating the results acquired for virgin carbon fibres in the work at hand in the context of a publication by Morris et al. [46]. A simple extrapolation of the modulus-area-correlation of results obtained here would yield a modulus range of 257 - 287 GPa if the PANEX 35 fibres were only 2.5 μm in diameter, which is still far below the values reported [46]. It is thus deduced that the cross-sectional area is not the major factor to result in higher moduli but that precursor quality and the stretching process during carbonisation play a more important role.

A reasonable explanation [335] is that pores exist within the carbon fibre. Pores only increase the cross-sectional area of the fibre but have no positive effect in terms of the Young's modulus or tensile strength (cf. Figure 7.4). There is also reason to believe that the increase will be related to an increase in the orientation of the BSU [335]. The effect that smaller fibres show far higher values of the Young's modulus is particularly interesting in light of composite design. Here, smaller fibres would yield two benefits: firstly, each individual fibre possesses a higher Young's modulus. Secondly, the total fibre surface forming interfaces would be enlarged [234].

Data presented in literature revealed that smaller cross-sectional areas of carbon fibres correlate with higher Young's moduli [24, 58, 223]. A deduction of this correlation is, that manufacturers would benefit from optimised production processes aimed at increased stretching and reduction in cross-sectional area [7, 58, 146]. Marthur et al. [336] demonstrated that the Young's modulus of PAN precursors is increased by increasing the stretch ratio, resulting in a higher Young's modulus of the manufactured carbon fibres. Investments in process technology to increase the stretching and decrease the cross-sectional areas would pay off as the expected Young's moduli become higher and thus the fibres more valuable. The fibres investigated in the work at hand showed a wide range in cross-sectional areas (cf. Figures 6.1 and 6.2), resulting in Young's moduli ranging between 178 and 248 GPa (cf. Figure 6.5). Narrowing the peak of the distribution of cross-sectional areas would decrease the variation in fibre stiffness, allowing for more reliable composite predictions. Besides more reliable stretching procedures, improvement in precursor quality will be reflected in the mechanical properties of the carbon fibre [64–66]. One way is using very high molecular weight PAN as precursor [46]. Still, one goal of research and development will always remain the production of carbon fibres with the highest possible Young's modulus [17, 18]. Another goal should be the optimisation of process routes to produce fibres with highly repeatable properties, ideally at low costs.

Poisson's ratio of virgin fibres

Poisson's ratio was only investigated for virgin carbon fibres. It was demonstrated that the SiFiT setup and SiFiT procedure enables determining the Poisson's ratio of carbon fibres. Nonetheless, lateral contraction of thin fibres is hard to determine due to very small diameters [233]. As a consequence, for the results shown in Figure 6.10, the standard deviation is rather high. In addition to small diameters, several aspects are assumed to be responsible for this: the nominal accuracy of the LDS0200 is 0.01 µm. On average, the difference between the apparent diameter $d_{avg,0}$ on the lowest and highest load level is only approximately 0.025 µm and intermediate steps could not be investigated. In contrast to that, Krucinska and Stypka [171] were able to determine the Poisson's ratio for individual load levels.

By manually tracking the LDS position, it was tried to keep the focus in the same position of the fibre. However, this might not be accurate enough and the fibre itself might strain non-uniformly. To obtain the average of the apparent diameters $d_{avg,0}$, the oval cross-sectional area was transferred to a circular shape which might be a cause of errors. Other than that, with a median of 0.310 the result for the Poisson's ration obtained by the SiFiT procedure meets the expectations: values of 0.22 and 0.35 [233], 0.025 and 0.3 [234] as well as 0.26 and 0.28 [171] were reported in literature. The large variation in the scientific results underline the need for enhanced experimental procedures to determine the Poisson's ratio. The SiFiT setup proved to be capable of determining the Poisson's ratio on single fibre level. However, a laser diffraction sensor with higher accuracy would be recommendable to obtain more reliable results.

Tensile strength virgin and recycled carbon fibres

The failure of carbon fibres can be described by Griffith's theory [213, 214]. Surface flaws or voids, as for instance seen in the TEM investigations presented in Figure 7.4, are among the reasons for failure fibre [24, 64]. In addition, misaligned crystallites can initiate tensile failure. [215, 216].

Weibull's theory was frequently applied to calculated the tensile strength of carbon fibres [93, 137, 218, 219]. However, Thomason [220] doubted its applicability, arguing that weaker fibres fail prior to testing and thus causing a shift of experimental results towards higher tensile strength. To take this into account, a computer model was run. As a result of this preliminary study it as found, that a minimum of 35 single fibres should be tested.

Literature also suggests that a correlation between tensile strength and cross-sectional area exists, stating that tensile strength is higher for smaller cross-sectional areas [7]. While Fitzer [7] applied a quadratic fit to correlate tensile strength with fibre diameter, a selection of data from literature as presented in this work (Figure 2.10) suggested a linear relation.

Weibull-plots themselves (see Figure 6.7 and 6.8) do not reflect possible variations due to the cross-sectional area of the fibre. Thus, an additional representation of the tensile strength data is presented in Figure 7.5. As has been mentioned before, the cross-sectional areas of the PANEX 35 fibres varies largely despite the fact that the fibres tested here were all of the same type and brand.

Figure 7.5 displays all tensile tests for vCF and rCF for gauge lengths up to 20 mm. Tensile strength is plotted versus cross-sectional area, in this figure regardless of the gauge length.

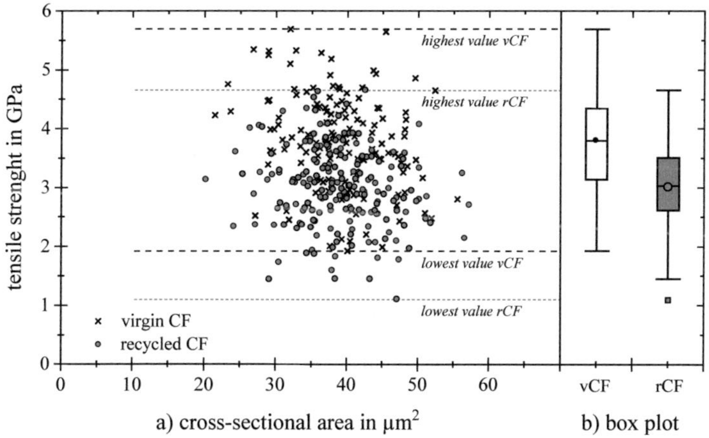

Figure 7.5: Tensile strength of carbon fibres, type PANEX 35; a) data points plotted as tensile strength vs. cross-sectional area to demonstrate the variation in both quantities. A correlation is not found. b) Box plot of the tensile strength for both fibre populations. An outlier is marked as grey square. Both visualisations suggest that the recycled fibres exhibit a lower tensile strength than virgin fibres.

The test volumes of the fibres with the lowest and highest cross-sectional areas (cf. Figure 2.10) differ by a factor of almost 3. Due to the volume effect, a correlation between fibre area and strength would be expected. However, the extensive distribution of the data points does not suggest a dependency between the tensile strength and the cross-sectional area for either vCF or rCF. Independent of geometrical investigations, there is a slight shift of the two point clouds as emphasised by the dotted lines and the box plots in Figure 7.5 b), representing the same data in another way. Particularly, the highest and lowest values are highlighted. Of 148 virgin fibres, 14 % have a tensile strength higher than the highest tensile strength measured for rCF, and 6 % of rCF do not reach the lowest tensile strength found for vCF. This finding does not correlate with the cross-sectional area of these fibres, but demonstrates the effect caused by recycling. This is in agreement with [106], where no significant correlation between the tenacity of individual PANEX 35 fibres and the linear density, a parameter linked to the fibre diameter, was found.

These results are in contrast to the collection of data reported in literature, presented in Figure 2.10 [24, 58, 223]. The fibres under investigation do only follow the assumption of the weakest-link-theory in terms of gauge length, not fibre volume (cf. Figure 6.8). Commonly, literature compares averaged values of strength and diameter of a fibre population [24, 58, 223] while in the work at hand, results of individual fibres are investigated. Equally, Jones and Duncan [33] correlated the strength of individual fibres to their diameters, concluding that increased diameters cause decreases in fibre strength. A linear trend is found, however, the widespread data points suggest a relatively low goodness of the fit. It is also stated that the effect is not observed for fibres with heat treatments during production below 1200 °C [33]. Yet, the inference that the missing correlation for the PANEX 35 fibres investigated here (cf. Figure 7.5) are due to low carbonisation temperatures, would be misleading.

Following the general discussion of fibre strength and its dependence on the cross-sectional area, the attention shall be drawn to the effect of recycling. In Figure 6.8, the tensile strength of virgin and recycled fibres was plotted versus the gauge length. On average, the recycling caused a reduction of the tensile strength of the PANEX 35 carbon fibres by approximately $21 \pm 6\,\%$. Despite some scatter, both, vCF and rCF showed a trend towards lower Weibull strengths for larger gauge lengths as proposed by literature [137, 228–231] and presented in Figure 2.11. The percental reduction in tensile strength of the recycled fibres found for different gauge lengths are also designated in Figure 6.8. Although the decrease varies for different gauge lengths, it can be found in all measurements. There is no evidence that the percental decline itself varies as a function of gauge length.

A comparison of the tensile strength with results from literature is found in Figure 7.6. It compares the tensile strength of vCF and rCF, stated in Table 6.2, with values published in [137, 228–231].

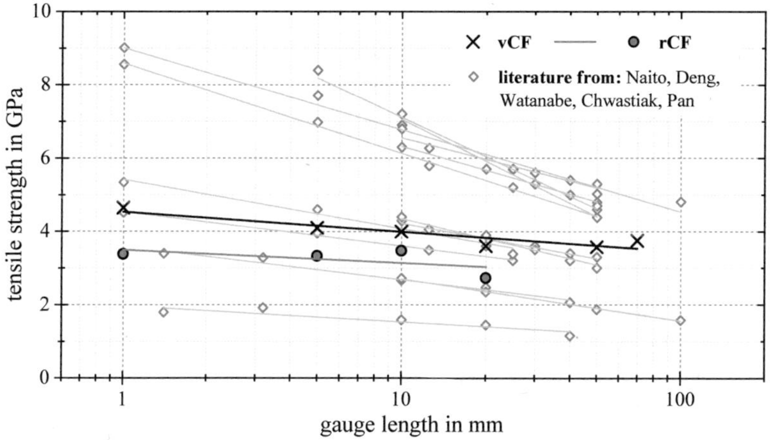

Figure 7.6: Experimental results of single fibre tensile strength compared to published data [137, 228–231]. Dark symbols present experimental data; open symbols and grey lines literature values.

Fibres investigated in the work at hand follow the expected trend of decreased tensile strength for increased gauge length. However, the slope of the curves is moderate compared to most of the data presented in literature. As stated before, different publications concerning single fibre properties of virgin and recycled carbon fibres are difficult to compare as test setups, procedures, fibre types etc. may vary significantly.

Without claim to completeness, Table 7.2 presents the results of various studies concerned with single fibre strength of recycled fibres compared to their virgin counterpart. In addition to pyrolysis as applied in the work at hand, some alternative fibre reclamation techniques have been reported. Besides the process, Table 7.2 lists the single fibre tensile strength of virgin fibres and the loss due to recycling. In this context, "loss" is the opposite of "retention", a term used to describe the remaining strength after recycling. In case a range is given, the publication was concerned with a parameter study and the highest and lowest loss are summarised here. Column "evaluation" lists whether tensile strength was evaluated according to Weibull's theory (wei) or if an average value was calculated (avg). In addition, reference and year of publication are included.

Table 7.2: Literature review: impact of recycling on single fibre tensile strength. The columns state the recycling route; the tensile strength of the virgin fibres and the reported loss due to recycling; whether strength was evaluated by Weibull's theory (wei) or by calculating the average value (avg); the reference and the year of publication. The first line represents the results found in the work at hand (cf. Table 6.2). A range of values for cited publications indicates that different recycling parameters were studied and lead to a variance in the tensile strength. Details are found in the publications.

recycling route	**tensile strength**		**evaluation**	**reference**	**year**
	virgin CF in GPa	loss in %			
pyrolysis	3.6 - 4.7	13 - 27	wei	Table 6.2	2019
pyrolysis	3.7	3.9	wei	[337]	2009
pyrolysis	3.2	10.3	avg	[338]	2011
pyrolysis	3.5	6.6 - 74.0	avg	[267]	2011
pyrolysis	5.9	13.4 - 70.2	avg	[312]	2012
pyrolysis	5.0	0 - 83.1	wei	[84]	2012
pyrolysis	3.4	11.8 - 29.4	avg	[311]	2018
microwave pyrolysis	4.1	20.3	avg	[339]	2004
fluidised bed	3.9	26.7	wei	[340]	2002
supercritical water	4.1	2.2 - 11.0	wei	[307]	2008
chemical recycling	4.1	0.4 - 15.9	wei	[308]	2008
supercritical propanol	4.1	4.6	wei	[341]	2006
supercritical propanol	4.3	0.3	wei	[332]	2006
supercritical propanol	5.2	0.3	wei	[332]	2006
supercritical propanol	4.3	2.1	wei	[332]	2006

Still, the comparison can be ambiguous due to the several reasons: some of the recycling parameters are corporate secrets; gauge lengths are not always provided and the statistical evaluation procedures vary. As described earlier, the experimental testing procedure, the type of fibre clamping, the free length between the clamps and the determination of the cross-sectional area are crucial parameters in single fibre testing. For instance, volume effects must be considered for different gauge lengths and diameter assumptions will be better for almost circular fibres than for oval or kidney-shaped fibres. Results of tensile strength are also depending on the evaluation procedure; Weibull strength and average strength are not directly comparable. Still, several parameters were not provided in some of the publications and cannot be taken into consideration here. Therefore, Table 7.2 is only supposed to provide a general overview of the impact of recycling on single fibre strength. Results collected in Table 7.2 do not provide information about multiple recycling runs of the same fibres.

Despite the described difficulties in comparing single fibres that have undergone different recycling procedures with various parameters, some trends can be deduced from Table 7.2: tensile strength can be significantly reduced during the recycling. Great care has to be taken to optimise recycling parameters and to guarantee a certain fibre quality for later re-use in composite production [281]. Some publications reported a loss of fibre strength below 5 % [84, 307, 308, 332, 337, 341]. On contrary, strength reductions of up to 75 % must be expected [84, 267] in case recycling parameters are not optimised. The tensile strength after recycling varies between different publications but also within single publications when a parameter study was run. For instance, Pimenta and Pinho [84] observed a variation in the loss of tensile strength from 0 to 83 % and Nahil and Williams [267] from 7 to 74 %. Commonly, it has to be decided between two opponent goals: mild recycling conditions promote better mechanical properties at the expense of more resin residue; harsher recycling conditions lead to a better removal of resin but can cause severe fibre damage [84].

The range observed in the work at hand lies well in the range found in literature. Due to the aforementioned difficulties in relating the results of different works, the most reasonable comparison of the properties observed in the work at hand is with publications [84] and [337]. Both researched pyrolysis routes and applied Weibull's theory for determining strength. Due to the large variation, all that can be said is that the results found here compare well with the observations in [84, 337].
The difficulties in comparing properties of recycled fibres are transferable to industry and are of commercial interest: the recycled material available on the market is commonly not declared and the users must deal with mixed material feedstock of unknown properties [84]. It is expected that commercial companies like *carboNXT GmbH* have optimised their recycling procedures to induce minimal damage in the fibres.

Summarising tensile properties, the following aspects should be mentioned: The Young's modulus showed a dependency on the cross-sectional area but no variance as a function of gauge length was found. As expected, recycling had no effect on the Young's modulus. In contrast, tensile strength was significantly reduced for fibres recycled by means of pyrolysis. Tensile strength correlated inversely with measuring length but no relation to the cross-sectional area was observed.

7.2 Discussion of surface and fibre/matrix-interface properties

7.2.1 Surface investigations with AFM and SEM

As stated earlier, results for the micro scale roughness parameters R_{max}, R_z and R_a indicated no significant influence of recycling on the waviness of the fibres. The typical lengthwise crenellations are clearly visible in SEM and AFM images (cf. Figure 6.11 and 6.12).

Results of different micro roughness parameters, presented in Table 6.3, lie well within the range reported in literature [315–317]. A comparison is given in Table 7.3.

Table 7.3: Surface roughness of carbon fibres as found in literature compared to experimental results (vCF, rCF). Rows 1 and 2 correspond to the results presented in Table 6.3. The 1st and 2nd column quotes the information concerning fibre material and sizing as stated in the corresponding publications.

material		maximum roughness R_{max} in nm		mean roughness R_a in n		reference
		mean	SD	mean	SD	
PANEX 35	vCF (sized)	224	68	45	15	–
PANEX 35	rCF (pyrolysed)	200	50	40	12	
CF (AS-4)	unsized	49	–	8	–	[315]
CF (AS-4)	PTPO sized	86	–	45	–	
CF (PAN)	sized	89	–	5	–	[316]
CF (PAN)	sized (modified)	107	–	9	–	
CF	commercial	210	–	37	–	[317]
CF	desized	218	–	41	–	
CF	0.7% DGEBA	161	–	34	–	

As this literature review reveals, sizing can have an influence on the roughness values of carbon fibres. However, no significant influence was found comparing vCF and rCF in the work at hand. Besides changes in the roughness values due to recycling, the question whether the surface properties affect measurements by laser diffraction is of interest in the work at hand: as reflected in Section 3.3, fibre roughness reaching peak roughness values R_{max} of up to 250 nm might impede the investigation by laser diffractometry, overestimating the cross-sectional area up to 14 %. However, in reality, the measuring errors due to the "smoothening effect" will be far less, as discussed in detail in Section 3.3.2 (cf. Figure 3.35). In this context, it is worth stating that no difference in laser diffraction measurement between virgin and recycled fibres are to be expected due to the likelihood of the roughness parameters. Yet, it was not possible to experimentally measure the roughness of the very same fibres used for mechanical testing. Measuring errors due to the deviation from a strictly convex shape cannot be evaluated in-situ. This microscale non-convexity was found to be not particularly pronounced for fibres of type PANEX 35. Nonetheless, about 37 % of the fibres showed minor non-convexity (cf. Figure 6.3). Even though the "crenelated" structure of carbon fibres might influence laser diffraction measurements, one should bear in mind that all other experimental procedures recommended in BS-ISO 11567 [1] will be likewise prone to roughness-related measurement errors. This is particularly true for side views in optical or electron microscopy, as undercuts are hidden behind protruding features. Vibroscopic methods are not assumed to be significantly affected by surface roughness but are, according to standard DIN-EN-ISO 1793 [155], not recommended for fibres possessing high stiffness.

Notwithstanding the fact that the roughness parameters are similar for vCF and rCF, effects of recycling can be likewise recognised in SEM and AFM investigation. The lengthwise crenellations are clearly visible on all carbon fibres under investigation but are more pronounced for recycled fibres. Compared to vCF, rCF exhibits other features on the surface, presumably residues of the polymeric matrix that have not been entirely pyrolysed. Overall, the nanoscale surface roughness appears to be increased, an observation that coincides with SEM investigations. Summarising, Figures 6.11 and 6.12 undoubtedly demonstrate that the surface structures of carbon fibres can be affected by the pyrolysis of a composite.

AFM investigations in [342] found that the roughness of sized PANEX 35 carbon fibres was 30 % less than that of oxidised fibres. Nanoporosity is resumed in [320] and correlations of the nanostructure of carbon fibres with their mechanical properties were discussed in [319]: it was concluded that tensile

strength decreased for fibres with increased aspect ratio of pores (length/width of the pore) and the effective pore area. The recycling-induced reduction in tensile strength found in the work at hand (cf. Figure 6.8) could be attributed to similar effects.
Surface-near pores, like those in Figure 7.4, are believed to have an influence on tensile strength [24, 65], regardless of the fact whether the fibres are virgin or recycled.
An increase in surface roughness might be an indicator of fibre damage. The reduced single fibre properties in terms of tensile strength are assumed to be caused by the structural changes of the surface. Possible reasons for the reduced tensile strength are surface defects and the increase in nano-scale surface features.

7.2.2 Chemical properties of the fibre surface

As summarised in [343], virgin and untreated carbon fibres usually possess surface areas, measured by BET analysis, of less than $1\,m^2\,g^{-1}$. Microporosity and surface area are reduced by applying higher manufacturing temperatures [343, 344]. With a surface area of $1.64\,m^2\,g^{-1}$ the virgin and sized PANEX 35 fibres studied here showed slightly higher values. Stoeffler et al. [300] determined a surface area of $1\,m^2\,g^{-1}$ for PANEX 35 by N_2-adsorption. Surface area cannot only be significantly increased by surface treatments [343] but also by recycling of carbon fibres, commonly intensified by subsequent activation processes [267, 300]. For instance, Stoeffler et a. [300] found surface areas of 18 and $42\,m^2\,g^{-1}$ for recycled carbon fibres. Yet, a comparison was hardly possible as the examined fibre waste was of unknown origin and specification. In the work at hand, the surface are was approximately doubled to $3.15\,m^2\,g^{-1}$. This finding is consistent with the roughness in AFM investigations, Figure 6.12.
Carbon fibres type PANEX 35 were investigated by inverse gas chromatography (iGC) in several studies and oxidisation and sizing effects were dis-

cussed [106, 345–347]. Hereinafter, those findings shall be compared to the experimental results on virgin (vCF) and recycled fibres (rCF) obtained within the work at hand. Therefore, data published by Huson et al. [106], Li et al. [345] and Kafi et al. [346, 347] is reprinted together with the experimentally determined surface energies in Figures 7.7, 7.8 and 7.9. For the reference labelled "Kafi2013" [347], only data for infinite dilution was available (symbols). For all other references, surface energies are plotted as a function of fractional surface coverage.

Sizing reduces the heterogeneity of carbon fibre surfaces, an aspect that was also observed when comparing vCH (sized) with rCF in the work at hand [106]. Considering the total surface energy γ_s^T the experimental results are akin to the respective literature: the virgin fibres vCF have been sized ex-works with an epoxy sizing, and γ_s^T concurs with the range found in literature. The total surface energy of the recycled fibres (rCF) is in the range of data published on oxidised fibres.

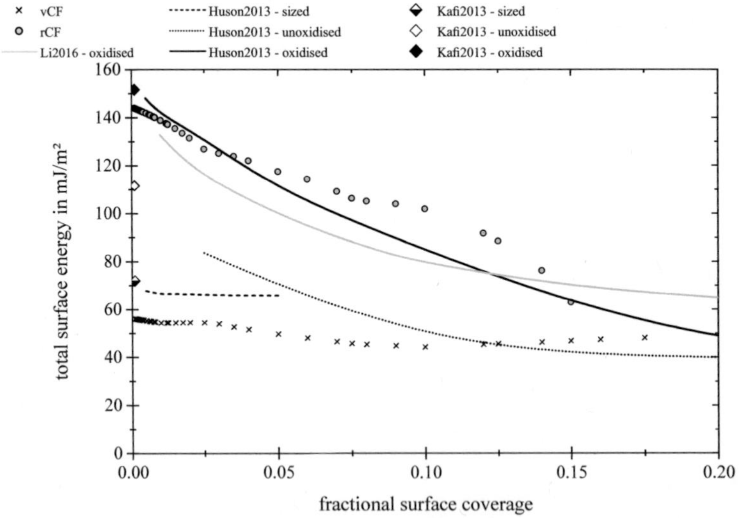

Figure 7.7: Comparison of total surface energy γ_s^T as a function of fractional surface coverage of experimental data (rCF and vCF) with respective literature [106, 345, 347].

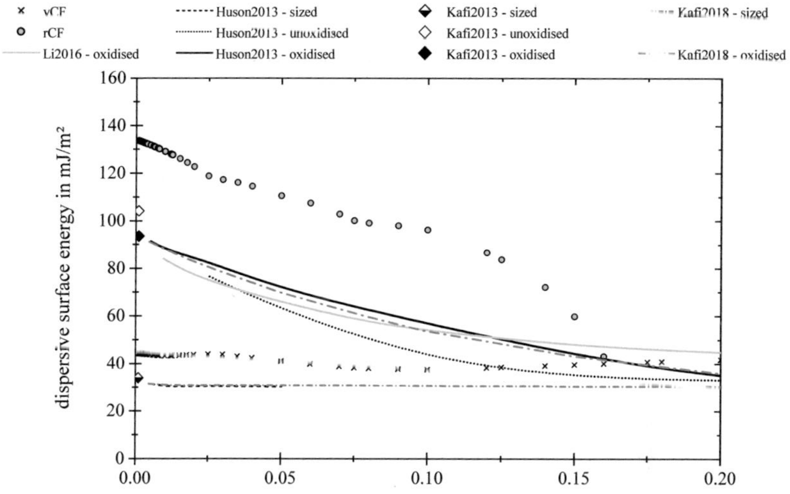

Figure 7.8: Comparison of dispersive surface energy γ_s^D as a function of fractional surface coverage of experimental data (rCF and vCF) with respective literature [106, 345–347].

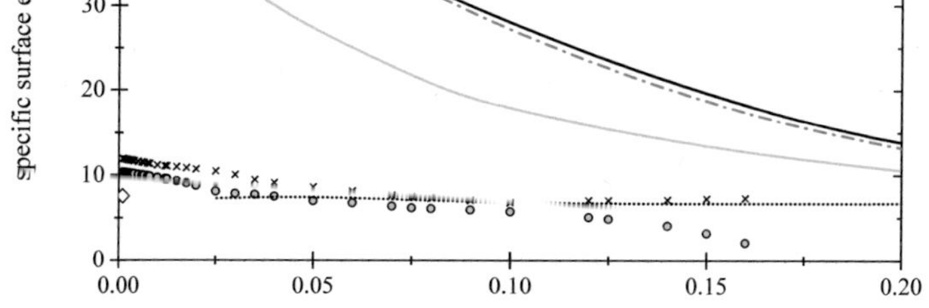

Figure 7.9: Comparison of specific surface energy γ_s^{AB} as a function of fractional surface coverage of experimental data (rCF and vCF) with respective literature [106, 345–347]. *The legend is found in Figure 7.8.*

The recycled fibres were neither re-sized nor post-treated subsequent to pyrolysis. Considering the dispersive surface energy γ_s^D and the specific surface energy γ_s^{AB}, the experimental results for vCF again meet the expectations derived from published data. Yet, the share of γ_s^D is unexpectedly high while the share of γ_s^{AB} is low compared to results published for oxidised PANEX 35 fibres. Results of iGC show that sized fibre (vCF) are energetically homogeneous, while the recycled fibres (rCF) show high degrees of heterogeneity. This indicates a complete coverage of the virgin fibres by sizing [106].
In view of the materials investigated in the work at hand, the following conclusion is drawn: The formation of new pores can be linked to the occurrence of increased nanoroughness seen in AFM investigation and with the decrease in tensile strength of the recycled fibres due to an increased amount of micro-defects on the surface. The correlation of tensile strength and flaws was described in [64]. Producing carbon fibres based on PAN fibres spun under clean room condition and thus with reduced amount of flaws lead to an increase of carbon fibre tensile strength by 80 %. It was further found in SEM investigations that failure was initiated by surface flaws [64]. It can thus be deduced that an increase of flaws on the surface due to recycling will result in reduced tensile strengths.

7.2.3 Interfacial properties: fibre/matrix-adhesion

Pull-out and push-out test show several disadvantages. Particularly, the preparation techniques results in test specimens that might vary significantly from the actual composite. Further aspects are discussed in the following:
Pull-out tests can hardly mimic the real properties of the interface in a composite: they focus on a single drop of resin and a single fibre. Thus, the effective interaction between thousands of fibres present in a composite cannot be reconstructed by pull-out tests. Furthermore, the history of the material, that is the temperatures and pressures during processing, is not reflected in the interface properties. Specimen preparation in pull-out tests is time-consuming as the

thermoset drops need to be cured one by one. It must be assumed that a total of four valid tests for both fibre specification was not enough for statistic reliability. Another source of error is the input of an average apparent diameter. When applying the *Fimatest* standard procedure, the diameter is determined using a vibroscopic procedure. As argued earlier, this method is not reliable for non-circular carbon fibres [197]. It was thus decided in favour of an average diameter here. The epoxy matrix used for this experiment differs from the resin systems used in the mechanical test (HS, EP). The pull-out test can only serve as comparison between the two fibre origins and does not completely reveal the properties of the composite interface of the real HS and EP composites.

In push-out tests, interaction between matrix and several fibres are present as the thin sections are prepared from the composite itself. However, the conditions and the fibre-matrix-interactions might be altered during specimen preparation. Push-out tests are best applied to unidirectionally reinforced composites to ensure that mechanical loading by the indenter is aligned with the fibre directions [252]. Misalignments in push-out test were studied in [261, 262] with the actual result being better than expected. Still, measurement errors must be expected for push-out test on misaligned fibres. Evident from Figures 6.25, 6.26 and 7.13, the reinforcing fibres are not aligned in BMC materials and great care needs to be taken to execute push-out tests. Specific challenges found in the work at hand are displayed in Figure 7.10: Particularly in thin regions, a weak interface can cause premature failure (1), cracks along the interface, spanning several fibre widths, can occur (2) or a weak interface is indicated by detachments of the fibre from the matrix (3). Due to the random orientation of the fibres, a characteristic of BMC materials, fibres can be misaligned in regard to the testing direction (4). Fibres in direct contact with adjacent fibres should not be tested in push-out tests as the interactions are unpredictable (5).

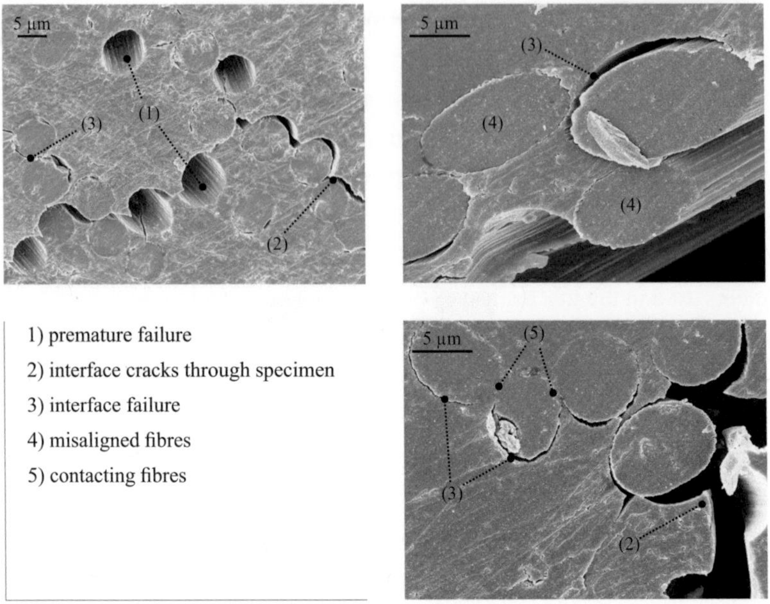

Figure 7.10: Push-out test run on BMC specimens revealed several challenges: 1) In thin regions of the specimen, fibres detached during the specimen preparation. 2) Cracks along the interface of several fibres occurred in the composite. 3) Weak bonding between fibre and matrix appeared, indicated by interface failure prior to testing. 4) Fibres not aligned with the testing orientation can occur due to the random fibre orientation of BMC materials. 5) Fibre in direct contact should not be considered in push-out tests as the interactions are unpredictable.

By carefully selecting suitable fibres, it was aimed at circumventing push-out tests on fibres subjected to the issues labelled 1 to 5. An advantageous characteristic of the push-out test is that specimens can be prepared from a composite panel, reflected the actual composite history and production.

In the work at hand, both tests were run in order to minimise the influence of the disadvantages of both procedures. In conclusion, neither pull-out nor push-out test showed a significant impact on the interfacial shear stress when replacing virgin fibres by recycled fibres (cf. Figure 6.17 and Table 6.4).

Still, the recycled fibres are tending to have a higher adhesion to the matrix.

The following can be hypothesised: as the apparent interfacial shear strength τ_{app} shows, the chemical bonding seems to be similar. As shown in Section 6.2, surface area and water uptake are higher for rCF, enabling more bonding between fibre and matrix. More importantly, the surface of recycled fibres shows higher nano-scale roughness (SEM, AFM), fostering frictional locking. An increase in frictional locking might be related to the increase in τ_f, τ_d and G_{ic} for rCF compared to vCF. As stated before, this can only be understood as a relative tendency between the two fibre origins.
The increase in surface area (determined by BET-measurements) and increase of surface energy (by iGC) might reflect in the friction related interface properties investigated mechanically, but did not cause an increase in interfacial shear stress.
Jiang et al. [332] compared the single fibre interface properties by means of microbond tests. The tested fibres of type T600S, T700S and STS5631 were combined with microdroplets consisting of *Hexcel Ltd.* epoxy resin *DW2*. Only interfacial shear stresses are reported, ranging from approximately 60 to 80 MPa. Recycling the fibres caused a minor but insignificant decrease in the interfacial shear strength [332]. Both, the absolute values and the effect of recycling, are consistent with the results presented in Sections 6.2.4 and 6.2.5.

One should not necessarily assume that variations of the interface by recycling impairs the composite properties. Positive effects could occur thanks to frictional locking between matrix and fibre or chemical activation of the fibre surfaces. Stated in [315, 320], an increased roughness of carbon fibres can foster interlocking effects. The increase in friction-related interface properties obtained in pull-out tests with recycled fibres, presented in Section 5.4.1, can be related to the occurrence of more distinct surface features (cf. Figure 6.12).

7.3 Discussion of composite properties

7.3.1 Process-related material characteristics

As expected for a composite with random orientation of carbon fibre bundles, results of density measurement and thermogravimetric analysis show scatter and inhomogeneities in the results. Specimen sizes are rather small and the fibre/matrix-ratio depends on the position of the sampling.

Density was determined using gas pycnometry. A comprehensive study by Rude et al. [324] proved the applicability of helium pycnometry for glass and carbon fibres as good agreement between the results and comparable literature was found. In the work at hand, the mean density of the recycled fibres was 5.6 % higher than that of virgin fibres (cf. Table 6.6). For vCF, the measured density is equal to the data reported by the supplier (Table 2.1). The increase of density for the recycled carbon fibres was not investigated in detail but might be due to the loss of polymer sizing on the fibre. For instance, Petersen et al. [348] detected that removing the fibre sizing by soxhlet extraction or pyrolysis caused an increase in density of glass fibres of 1 to 6 % . In [324], carbon fibre samples of type *IM7* and *T300* were reclaimed by acid digestion of laminates. Thereby, the sizing is presumably removed. Still, the values reported in [324] were in the range of data provided in other sources, but not increased. However, comparability is difficult as measurements on sized and unsized fibres were not run in the same laboratory and with the same method.
It shall be noted that the density was determined for quality assessment and could not taken into account during the BMC production presented in Section 4.3. There, values given by the producer were used for determining the mixing ratios and differences between recycled and virgin fibres were not considered. In density measurements, the sizing is considered as fibre as it also fills a certain volume. On contrary and contingent on the fact that the sizing is pyrolysed together with the polymer, it counts as matrix in thermogravimetric

measurements. This fact might be the reason for slightly higher (approximately 5 %) mean fibre weight content of the recycled materials compared to the composites containing virgin fibres (Figure 6.19). In the former, the sizing was already removed during the recycling process, while for the latter, the sizing is decomposed during TGA measurements itself.

Nevertheless, the results of DMA, TGA and density measurements (cf. Section 6.3.1) suggest that the quality of the BMC panels is sufficient for mechanical analysis. All three investigations confirmed that all panels bear sufficient resemblance to be compared directly. The results of the mechanical testing and effects of recycling are discussed in the following.

7.3.2 Comparison between bulk and sheet moulding compounds

The similarities and differences between BMC and SMC processes and products were described in Section 2.2.2. Given the process schemes presented in Figures 2.13 and 2.14, it can be derived that it is easier to apply recycled and waste fibres in the BMC process. It was demonstrated in the work at hand that patches or recycled fibres can readily replace virgin fibres cut from a roving. The kneading step ensures a thorough mixture of fibre and resin (cf. Figure 6.28). On contrary, standard SMC routes rely on new rovings as feedstock for the cutting unit (Figures 2.13). To use recycled materials in a SMC route, the conventional cutting unit would have to be replaced by machinery that is capable of dosing shredded fibres homogeneously. It is expected that this additional step is costly. As the kneading step is omitted in SMC, the material might not be homogenised sufficiently.

In this context, it is of interest to what extent BMC materials can replace SMC materials. In [150], the author studied this issue by comparing BMC with dry fabric waste fibres (PANEX 35) with SMC made with the same fibres from rovings. Resin HS served as matrix. It was found that both materials are akin in their structure: both contain fibre bundles rather than single fibres. More importantly, BMC was competitive in terms of mechanical properties. Flexural and tensile strength, flexural and tensile modulus and Charpy impact test were in the same range for BMC and SMC but again with large scatter in the mechanical properties. As described before [349, 350], the scatter in the mechanical properties is rather a consequence of the random bundle orientation than of the recycling step.

In summary, BMC routes offer great potential for re-using recycled fibres in composite structures. In terms of their mechanical properties, they can compete with SMC parts based on the same ingredients and from a perspective of sustainability and economy, they even might outrun their SMC analogue thanks to reduced costs for machinery (BMC process) and material (recycled fibres). Approaches to integrate the direct BMC process in a holistic recycling approach are outlined in Chapter 8.

7.3.3 Microstructure and failure of bulk moulding compounds

In the BMC process, the fibres were added in bundles. The BMC kneading process was designed to unbundle the rectangular fabric patches and to impregnate them without separating bundles into individual fibres. As a consequence, the bundle-like characteristics are typical for the direct BMC process [150, 251]. Whilst no experiments to count the amount of fibres per bundle in the composites were performed, SEM studies indicated the existence of bundle structures within the composite. The bundle-like structure is displayed in Figures 6.25, 6.26 and 6.27 as well as in Figure 7.11 for material combination vHS.

The same bundle-like characteristics were also made visible by investigating microsections (Figure 6.28) by light microscopy. As adjacent fibres within a bundle (labels 1 to 11) show similar shapes, it can be concluded that they are orientated in the approximately identically direction. In turn, the main orientation of neighbouring bundles can vary significantly.

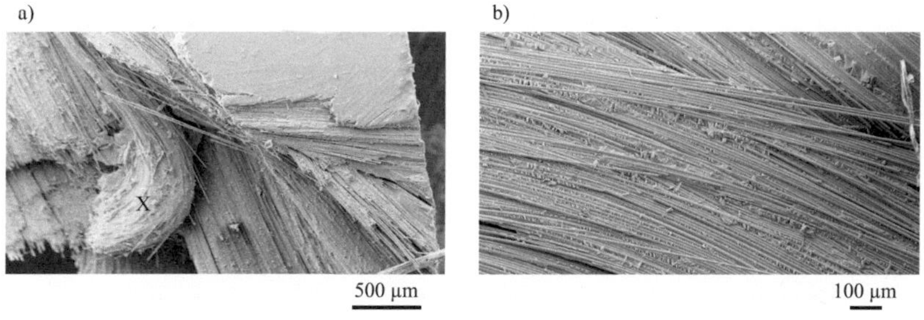

Figure 7.11: Fibre bundles in a BMC composite (vHS). Hundreds of fibres within a bundle run almost parallel and follow the same curvature, which can be distinctively seen in bundle X (a).// *Images taken with SEM/IAM*

It can be assumed that the more elongated the oval fibre shapes appear, the larger the angular deflection between optical axis and fibre direction (for instance Figure 6.28, bundles 3 and 10). Bundle 10 is of particular interest as it indicates that the curvature is located right in the sectional plane. The majority of fibres is clustered in bundles and only few are present as single fibres. At intersections of fibre bundles or intersections of bundles and stitching yarns, regions rich in matrix can occur. Most importantly, a thorough impregnation of fibre bundles by the matrix can be seen. Most fibres are impregnated and even in case individual fibres are in direct contact, the interstices are filled with matrix. In addition to the occurrence of fibre bundles, BMC structures contain randomly orientated fibres, which are a cause of the quasi-isotropic behaviour. Despite the fact that recycled fibres were less attached to each other, clusters appeared in the composite panels.

SEM investigations and fractography

The non-uniform behaviour can be visualised with different experimental tech niques: fracture surfaces investigated in SEM (Figures 6.25, 6.26 and 6.27) show the bundle-like character and the random orientation of fibre bundles. Furthermore, it can be clearly seen that fibre bundles not oriented in loading direction tend to be detached from matrix regions, which means that the high tensile strength of the fibres cannot positively affect the strength of the composite. Particularly, the failure behaviour becomes visible in SEM when working with increased magnifications: Figure 6.27 displays different matrix and fibre combinations. Once more, the bundle characteristic of BMC is found. A prominent feature found in all images are imprints ("negatives") of the fibres in the resin. The fibres have been delaminated from the matrix during the tensile test, leaving behind distinctive imprints. Together with the fact that the fibres show almost no residue of matrix, it is concluded that the interface properties are weak, particularly when using HS resin. Since statistically speaking, areas with transverse fibres always occur in the parallel region of the specimens, the tensile strength of the composite cannot be expected to be higher than the strength of the unreinforced matrix. BMC composites containing virgin or recycled fibre display another difference: when using rCF, the stitching yarn had been eliminated during pyrolysis whereas when using virgin carbon fibre patches, the stitching yarn was still in place. Remains of stitching yarn were frequently found on fracture surfaces of virgin fibre composites (for instance in Figure 6.25c)). As they might cause the appearance of polymer-rich regions, they facilitate crack initiation. This phenomena was summarised by Greenhalgh [351] for composites based on non-crimp fabrics. Resin rich regions and the remains of stitching yarn might not only weaken the material but also induce secondary damage. Stitching yarn and the associated regions of high resin content were for instance found in Figures 6.25 and 6.28. Eliminating the stitching yarn prior to the production of vCF-composites fibre was considered in the work at hand: removal of the stitching yarn by laser or chemical treatment was feasible. However, as such treatments might induce unpredictable fibre damage or change the composite properties, no detailed analysis of these materials was carried out. Another effect of stitching yarn on composite failure, the formation of voids in the vicinity of the yarn, was not observed on any of the failed BMC specimens. Such failure is more likely to occur in traditional stitched composites, based on preforms with highly aligned fibres [351].

Digital image correlation to visualise failure behaviour

Inhomogeneous distributions of local strains also become visible when evaluating data recorded by digital image correlation during the tensile test. In the exemplary Figure 7.12, the region of interest is displayed as strain maps. The region of interest was set as a rectangular region of 60 × 16 mm in the centre of bone-shaped specimen. As can be seen, the strain is distributed inhomogeneously and local strains significantly exceed the average strain calculated for the region of interest, graphed in the corresponding stress-strain-curve.

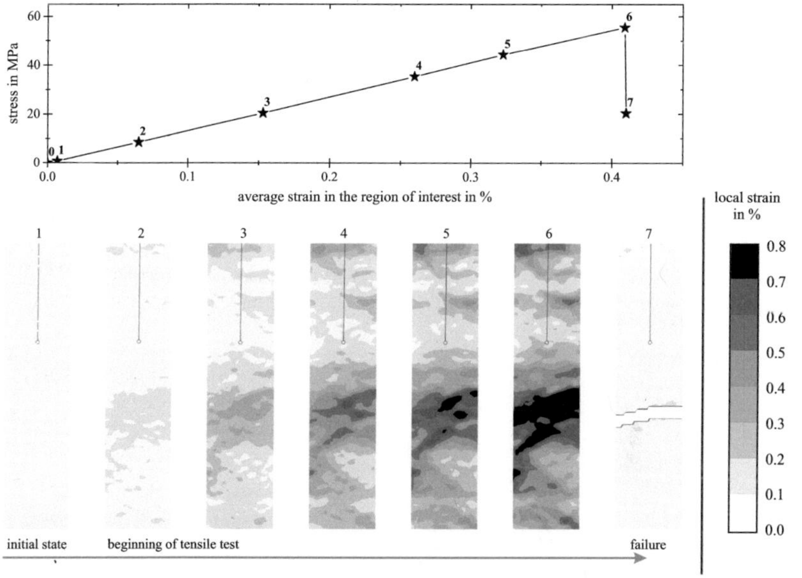

Figure 7.12: Exemplary stress-strain curve for an epoxy-BMC. The local strain is shown as strain map derived from digital image correlation.

Where fibre bundles are poorly aligned in respect to the loading direction, the fibres cannot fully contribute to the specimen reinforcement and the local stiffness is low; the specimen deforms more distinctively. Region of high strains are randomly distributed in the specimen and are related to the weak spots described before. Failure is likely to be initiated in regions with high local strains, which frequently appear where fibre bundles are not aligned in tensile direction or where a resin-rich region exists. Nevertheless, it should be noted that the specimens were only observed with DIC on one side, meaning that weak spots on the averted side might remain undetected.

Computed tomography as visualisation tool for BMC behaviour

Another complementary technology to reason that misaligned fibre bundles are the relevant factor for BMC failure are CT scans. CT scans of random samples were acquired using a YXLON *precision micro CT* with a tungsten target. Some specimens have been scanned subsequent to tensile testing. Distinct features of the BMC material are well described with the help of Figure 7.13. The top view of the specimen shows the bundle-characteristics of the BMC: some bundles are aligned in the testing direction, some rather transverse to it. A crack is seen adjacent to a transversely orientated fibre bundle as this appears to be a weak spot in the specimen. It should also be emphasised that regions rich in resin and region with a high local fibre content are seen. A finding corresponding well to the TGA results presented in Figure 6.20, where local variations of the fibre weight content were found.

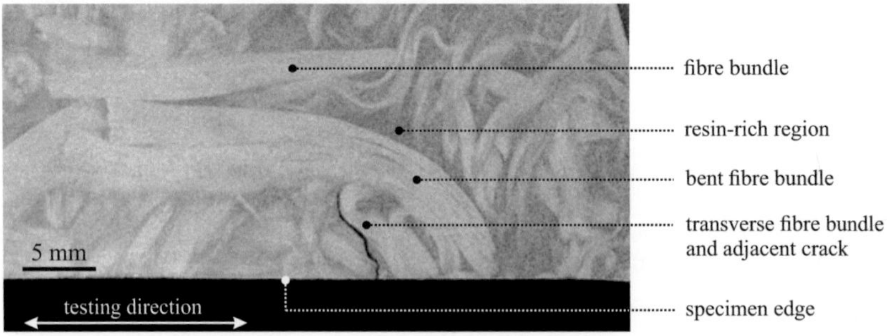

Figure 7.13: Computer tomograph scan of an internal plane in a BMC specimen, top view.

Summarising discussion of the failure mechanism of BMC

The fracture behaviour can be described according to the observation published by Hughes [234], presented in Figure 2.16. Traverse fibre splitting was not observed, neither through the core (d) nor surface-near (c). Instead, failure is mostly apparent at the fibre/matrix-interface. Frequently, the fibre surface appears to be clean with only minor traces of resin residues attached to it. It remains unclear whether the residues seen in SEM are attached to the fibres or if they consist of particles which remained as impurities on the fracture surface. As SEM images were taken post mortem, they do rather demonstrate crack propagation than initiation [234, 351].

As weak spots and their distribution dominate failure, the scatter is increased and material characteristics and differences between virgin and recycled fibres loose their significance. On the one hand, this proves beneficial in terms of using recycled carbon fibres in composites with randomly-orientated fibres. On the other hand, alternative procedures to evaluate the tensile behaviour of the four materials investigated here, are needed.
Given the described structure of bulk moulding compounds, a decrease of the fibre properties due to recycling should have the same effect in both resin systems used in the work at hand. As aforementioned, bulk moulding compounds consist of randomly-orientated fibre bundles. Most likely, failure is initiated at weak spots within the material. The most common weak spots to be expected in BMC are resin-rich regions or fibre bundles mostly perpendicular to the testing direction. Both are unavoidable consequences of the manufacturing process but not a result of fibre reclamation. Anisotropic material behaviour is a characteristic of SMC and BMC materials [349, 350]. It is believed to cause distinct scatter in material properties. Computed tomography (CT) and DIC emphasised this finding and allowed a correlation between bundle orientation, increased local strains and final failure of the materials. Concerning these observations, no significant differences were seen due to the state of

the fibre: failure mechanisms in these material systems are dominated by the behaviour of the matrix. While standard tensile tests showed no difference between both resin systems, bulk moulding compounds evaluated according to a Weibull procedure revealed that Weibull strength was approximately 60 % higher for epoxy based materials. Recycling had no significant effect on the Weibull strength of the tested composites.

Summarising the tensile tests and the failure mechanism of the bulk moulding compounds, it can be concluded that failure occurs where fibres and fibres bundles are not aligned with respect to the loading direction and/or where regions rich of resin are present. There, cracks propagate through the matrix or the interface along the misaligned fibre bundles. In the BMC panels presented in the work at hand, the fibres themselves play a minor role for the failure mechanism. It is thus unpertinent for the BMC process introduced in this work, whether virgin or recycled fibres are used.

7.3.4 Non-linear strain behaviour of composites

Carbon fibres [2] and their composites [205, 208, 209] show non-Hookean behaviour, with Young's modulus increasing with increased strains. The effect is predominantly seen in single fibres and in polymers reinfored with unidirectional carbon fibres [205]. As expected, no relevant non-Hookean effect was seen in the BMC materials investigated in the work at hand. Due to the random orientation, only a minor fraction of fibres is aligned in the loading direction. The non-Hookean effect is believed to be caused be reorientation of crystallites in the fibre occurring when the fibre is loaded in tensile direction [2]. As a consequence, only a fraction of the total fibre volume content would show non-linear behaviour in BMC parts, whereas in unidirectionally reinforced composites, the non-linear behaviour occurs [205, 209, 211, 212]. In composites reinforced with ramdomly orientated fibres, shear and matrix deformation within the composite superimpose possible stiffening effects of the fibres [209]. The stress-strain-diagrams of the BMC composites, for instance shown in Figures 6.22 and 7.12, offer no indication of a non-Hookean behaviour.

7.4 The effect of using recycled fibres in composites with random fibre orientations

So far, the potential of a direct BMC route for using non-impregnated fibre waste and pyrolysed fibres was examined. Material properties were compared on fibre, interface and composite level. Hereinafter, the experimental observations are put into context with the state of research in carbon fibre and CFRP recycling. In the field of recycling, quantitative comparisons with the state of research are difficult to accomplish as too many parameters might differ between research projects. Still, some statements can be made: When evaluating recycling strategies for composites, cost and process complexity compete with mechanical performance of the final product. As a consequence, different procedures co-exist, depicted in Figure 2.19 [10, 84, 278, 281]. The direct BMC process introduced in the work at hand aims at coming to a compromise by manufacturing composites with intermediate properties in a low-cost, straightforward process.

With reference to Figure 2.19 it can be deduced that this BMC material neither competes with composites enriched with recycled fillers [288, 290–293], nor with those consisting of a highly orientated fibre architecture [289, 295, 299, 301, 302]. On the one hand, recycling by grinding composites or fibres inevitably leads to major down-grading of the expensive carbon fibres, whereas the target in the work at hand was to obtain long-fibre reinforced composites with fibre lengths of approximately 50 mm [150, 251]. On the other hand, a high degree of fibre orientation is beneficial in terms of mechanical properties but requires additional production steps, associated with increased costs and longer productions cycles.

A fair evaluation of the effect of using recycled fibres in BMC composites is only feasible in relation to composites with random long-fibre orientation. For instance, prepreg waste was used in [295] to manufacture BMC parts. Standard BMC containing virgin glass fibres were compared to the recycling-based carbon fibre BMC: the estimated material costs approximately tripled and reached 7 € per kg, however, the Young's modulus was also three times as high, reaching 30 GPa. This value is in the range of those found in the work at hand, however, direct comparability is not given as different materials were under investigation. Still, the work by [295] is motivated by the same goal as the direct BMC process introduced in the work at hand: to obtain well-performing composite at relatively low cost, competing with virgin glass fibre products while promoting sustainability by recycling. Turner et al. [295] focussed on prepreg waste whereas the work at hand deals with strategies to re-use dry fibre fabric waste and end-of-life parts. End-of-life parts were also used by Ogi et al. [298]. Without intermediate steps of fibre reclamation or alignment, CFRP scrap pieces were used to reinforce concrete. Due to the difference in material, no quantitative comparison to the work at hand is possible. Still, the work took a step ahead as the stage of fibre reclamation was successfully omitted. This strategy is worth considering for reinforcing polymers [298]. Feraboli et al. [352] compared panels produces by vacuum-assisted resin transfer moulding technology (VARTM). The virgin reference contained 62 vol.-% of twill carbon fabric. Fibres were recycled by a procedure based on acid digestion to produce panels with the same VARTM process, however, only achieving a fibre volume content of 33 %. The authors normalised their results to a intermediate content of 50 vol.-% of carbon fibres to show that the mechanical properties of virgin and recycled modifications were in good agreement and in the range of advanced SMC materials. The reliability of the interpolation was not further discussed and is thus debateable. In comparison with the work at hand, the modulus is in the same order of magnitude while it outperforms by a factor of about 3 in terms of tensile strength [352]. Interestingly, the appearance as well as the failure behaviour are akin to the BMC material presented here.

8 Application and outlook: holistic approach to BMC recycling

Presented in the state of research on composite recycling technology, Section 2.3, two main concepts of fibre re-usage are being researched [10]: firstly, production of intermediate products like wovens or hybrid yarns and reprocessing to high-performance composites. This concept aims on fibre alignment and orientation [289, 301, 302]. Secondly, waste treatment "as-received", thus using low-cost technology to make the end product economically attractive [150, 288, 291, 295, 298, 300]. These products have to compete with composites made with randomly-orientated virgin glass fibres. From today's perspective, both concepts can be used in parallel as they do not compete in the same market. On the contrary, both concepts can complement each other to achieve a closed-loop recycling of carbon fibre reinforced polymers.

A concept how to use the achievements of the work at hand in order to recycle multiple fractions of carbon fibre production waste will be presented in the following [251]:

As shown in this work, the properties of BMC materials are not affected by the use of recycled fibres. Failure in BMC is not dominated by the fibres. Instead, fracture occurs where fibre bundles are oriented unfavourably with respect to the load direction. Given the mechanical properties presented in the experimental part of this work, it is not feasible to use these materials to design load structures, for instance in automotive sector. However, this is not a consequence of fibre recycling, but an inherit disadvantage of BMC and SMC parts. On the other hand, BMC processes are expected to be cost-efficient while the usage of recycled fibres will not raise the costs [10]. As presented in this work,

apart from cutting recycling waste into smaller pieces and weighing, no additional processing steps are needed and the dry waste fibres are directly fed into the BMC process. This applies likewise for recycled fibres.
Despite the disadvantages of random fibre and bundle orientations and the accompanying scatter in properties, BMC appears to be a valuable technology when addressing carbon fibre recycling and dry fibre waste. A holistic approach how to reuse all fractions of dry fibre waste in the automotive sector was presented by the author in [251] and is shown in Figure 8.1

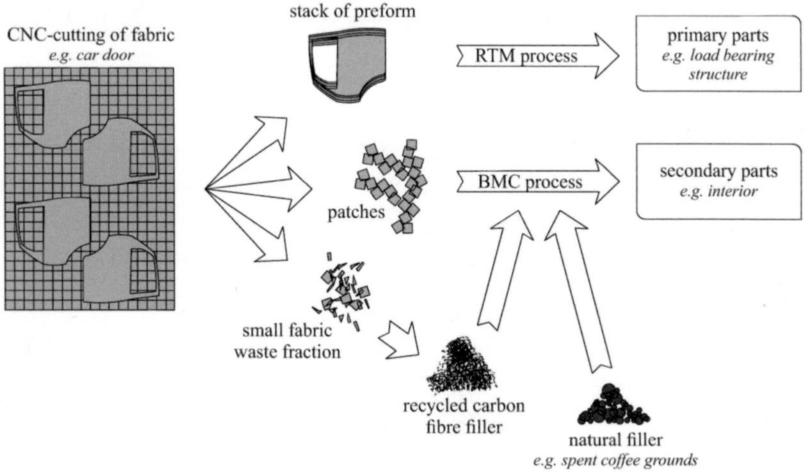

Figure 8.1: Flow chart of holistic recycling approach via a BMC route. Larger fractions of fabric waste are the reinforcement component in a BMC part, optionally, filler from recycled natural products (e.g. spent coffee ground) or recycled ground carbon fibres can be used [250].

In conventional processing routes, carbon fibre fabrics are CNC-cutted to a desired shape, for instance the door of a vehicle. While the plain fabric is rectangular, the part usually is not. As a consequence, excess material appears. This dry fibre fabric waste is cut into smaller pieces. All pieces (patches) larger than a certain threshold can be used as feedstock for the direct BMC kneading process described here. In addition, all fabric pieces below the threshold can be ground to carbon fibre filling material (CFF). The filler is blended with the resin and both are used in the BMC process. Carbon fibre fillers are already used and discussed [288]. Carbon fibre fillers are also available for purchase by some companies [1].

In this holistic approach, all carbon fibre waste can be reused to mould new components for the automotive sector and no material is disposed of, meeting the goals of European regulations [276, 277]. Given the results of the work at hand, rCF and rCF-based fillers can likewise be used. Even though this approach involves minor downgrading, the carbon fibres could serve as reinforcement in automotive applications, particularly in vehicle interior design or as semi-structural components. Components such as rear-shelves, cup holders, hood scoops or head-lamp housing are conceivable. BMC materials based on virgin glass fibres and calcium carbonate as filler have been used for decades in automotive applications like these [353, 354]. BMC components containing recycled carbon fibres or non-impregnated fibre waste offer an alternative, promoting sustainable car design. In terms of production, carbon fibres can be processed as easily as the glass fibres in the traditionally glass-fibre based BMC process.
It was further demonstrated that the filler of ground carbon fibre waste is beneficial in terms of lightweight design, compared to the standard filler calcium carbonate [251]. The carbon fibre filler made from ground fibres is not only lighter but has a stiffening effect thanks to a higher aspect ratio than conventional filling materials. Consequently, some of the long fibres can be replaced by carbon fibre filler. The material is compared to others in a so-called ASHBY-map, Figure 8.2. The concept is of particular interest for vehicle manufacturers who can recycle their production waste in-house to fulfil all regulations while producing carbon-fibre reinforced secondary parts [276, 277].

[1] https://www.car-fibertec.de/cf-flex, - 07.05.2019
https://www.haufler.com/content/carbonfaser/carbonfaserkurzschnitt/ - 07.05.2019

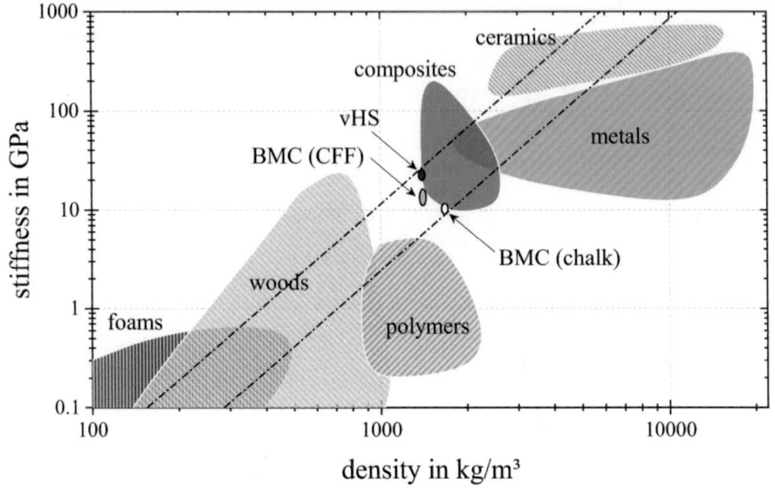

Figure 8.2: Ashby map with stiffness as a function of density. In terms of lightweight design, the carbon fibre filler "BMC (CFF)" is superior to calcium carbonate "BMC (chalk)" [245, 251, 280].

Here, the stiffness is plotted versus the material's density to provide information about the specific stiffness. SMC and BMC products or commonly used as panels and are thus loaded in bending. According to Ashby [245], the following material index M is then relevant:

$$M = \frac{E^{1/3}}{\rho} \tag{8.1}$$

The higher M, the better the material properties in terms of plate bending, as indicated by the dashed lines in Figure 8.2. All data points lying on one of the lines show the same ratio of stiffness to density. Being shifted upwards and to the left, the material with carbon fibre filler (CFF) are superior to the composites with the conventional filler calcium carbonate.

The philosophy behind the BMC process presented here, with or without fillers, is to circumvent the need of the complex processing steps needed to realign recycled carbon fibres. Instead of several costly and elaborate intermediate steps [289, 295, 301], the direct BMC process can use recycled materials while keeping fibre lengths significantly longer than for instance in injection moulding [300].

Given its versatility, the direct BMC process offers further possibilities in terms of holistic recycling procedures. In [250], the author investigated another hybrid recycling concept by combining waste carbon fibres with spent coffee ground as recycled filler, depicted in Figure 8.3.

Figure 8.3: BMC panels containing spent coffee ground offer opportunities for the interior design of sustainable future vehicles [250, 251].

In the automotive context, semi-structural components made with this material combination offer potential for applications in the interior of vehicles. In the works presented here, flat panels were produced. However, these BMC mixtures should exhibit the same versatility in terms of possible structures as standard BMC [239, 244]. The sustainable material containing spent coffee ground is expected to be low-cost and offers outstanding haptic and aesthetic properties. According to Ashby [245], nowadays a material not only needs to possess function, but also emotion. In this context, a material based on recycled fibres and a well-known waste product like spent coffee ground, might offer both in future applications.

As shown, the direct BMC process is a versatile process. Semi-structural parts containing recycled fibres can be produced. Optional, the material can be enriched with recycled filler, for instance recycled carbon fibre waste or spent coffee ground. It has been confirmed that both recycled fillers are superior to standard calcium carbonate in terms of lightweight design [250, 251]. The process offers opportunities for recycling applications without a need for intermediate steps such as fibre alignment or the production of non-woven fabrics. Returning to the boundary conditions presented in Section 2.3 [285], the direct BMC recycling process is compatible with today's recycling market, offers an technology to recycle both, composites and fabric waste, and is expected to be profitable due to low-cost operation. Nevertheless, like all other recycling strategies in the composite market, the process will have to prove its applicability to gain the industries interest and the confidence to make use of it [281].

9 Summary

When condensed to a minimum, this doctoral thesis is based on the following line of thought: increased demand for carbon fibre composites leads to an increased need to recycle composites without losing their valuable properties. Knowledge of the impact of recycling on carbon fibres and their composites is stringently required to convince manufacturers to place reliance on recycled materials. As a consequence, the work at hand was dedicated to the structure-property relationship of carbon fibre reinforced plastics and to what extent recycling has an impact. These correlations were investigated on all scales: from micro-level (single fibres), via meso-level (surface and interface) to macro-level (composite).

Single fibres testing was found to be a potent tool in investigating the recycling impact. However, state-of-the-art single fibre testing protocols were found to be immature, offering the potential for improvements.
Therefore, two scientific questions were in focus of this doctoral thesis and shall be briefly summarised hereinafter.
Firstly, how can single fibre testing procedures be improved? Particularly, how can strains be measured directly on the fibre and how can the non-circular shapes of carbon fibres be approximated?
Secondly, to what extent do the mechanical properties of virgin and recycled carbon fibres and their composites differ? Are recycling-induced changes of the single fibres reflected in bulk moulding compounds?

A literature review revealed that many types of carbon fibres are far from being circular in shape but rather oval or kidney-shaped. A fitting procedure based on assuming elliptical shapes proved to be useful in approximating the cross-sectional area of oval or slightly kidney-shaped fibres: results varied less than 10 % from the actual cross-sectional area, while state-of-the-art calculations based on circular assumption might deviate up to 50 %. A digital image correlation procedure to determine strains with the help of polymeric markers attached to the fibre was introduced. It was verified that results are not affected by the measuring technique itself and that the Young's modulus, if not determined directly on fibre level but in consideration of the machine's crosshead, might be underestimated by approximately 35 %.

The SiFiT setup and method, developed within the work at hand, proved to be a useful concept in determining the Young's modulus of carbon fibres. In addition, it provided irrefutable proof of the non-Hookean behaviour and the correlation between the modulus and the cross-sectional area of carbon fibres. None of these properties was affected by recycling. However, tensile strength was reduced by approximately 20 % for pyrolysed fibres compared to virgin carbon fibres. This might be due to the fact that the recycled fibres showed more distinct length-wise crenellations which are presumably correlating to an increased likelihood of surface flaws initiating failure. In addition, recycled fibres showed an increase in surface energies.

The interfacial shear stress showed neither a significant difference between the virgin and the recycled state, nor the applied testing methods pull-out and push-out.

In respect of composite properties, bulk moulding compounds were produced and tested. Two different types of resins (hybrid and epoxy) were combined with virgin and recycled carbon fibres. The mechanical behaviour and properties were similar, with an average stiffness of approximately 20 to 25 GPa. Epoxy-based composites showed a more abrupt failure behaviour than hybrid-resin composites, accompanied by an increase in Weibull strength of approximately 60 %, from 50 MPa to 80 MPa. The state of the fibre did not influence

the composite's Weibull strength. In terms of failure behaviour, all materials were resembling as failure of the composites could be attributed to misaligned fibre bundles as well as resin-rich regions. Both caused remarkable scatter of the mechanical properties. This failure behaviour is typical for BMC composites and was found for both, the virgin and the recycled fibres. Thus, it is not an effect of recycling but a generic attribute of BMC materials. Whilst this might be interpreted as a disadvantage of the BMC materials, the straightforward production concept is advantageous in respect of carbon fibre recycling.
It was further shown that a direct BMC route is equally suitable for processing virgin and recycled carbon fibres to produce quasi-isotropic composite panels. A holistic perspective on composite recycling in the automotive industry was suggested by adding recycled fillers, for instance ground carbon fibre waste or spent coffee ground.

A better understanding of the recycling impact on carbon fibre reinforced plastics will support the widespread usage of recycled fibres to fulfil legislative regulations of waste treatment and to encourage sustainable composite design. In this day and age, designing recyclable and sustainable products and components should be an honourable goal of researchers and engineers.

Bibliography

[1] British Standards: “BS-ISO 11567:2018: Carbon fibre - Determination of filament diameter and cross-sectional area: Carbon materials.”, **2018**

[2] G. J. Curtis, J. M. Milne and W. N. Reynolds: “Non-hookean behaviour of strong carbon fibres.” *Nature* Volume 220, 1024–1025, **1968**

[3] S. Das, J. Warren, D. West and S. M. Schexnayder: “Global carbon fiber composites supply chain competitiveness analysis.” *Clean Energy Manufacturing Analysis Center, Golden, CO, Technical Report No. ORNL/SR-2016/100—NREL/TP-6A50-66071* **2016**

[4] Zoltek Toray Group: “Technical datasheet - PX35 Multi-directional fabrics.” **2016**

[5] T. A. Edison: “US Patent 223,898 - Electric lamp.”, **1880**

[6] R. Bacon and C. T. Moses: “Carbon fibers, from light bulbs to outer space.” In K. G. S. Seymour, R B, editor, “High Performance Polymers: Their Origin and Development,” 341–353. Springer, New York, Amsterdam, London, **1986**

[7] E. Fitzer: “PAN-based carbon fibers—present state and trend of the technology from the viewpoint of possibilities and limits to influence and to control the fiber properties by the process parameters.” *Carbon* Volume 27, 621–645, **1989**

[8] B. A. Newcomb: "Processing, structure, and properties of carbon fibers." *Composites Part A: Applied Science and Manufacturing* Volume 91, 262–282, **2016**

[9] H. G. Chae, B. A. Newcomb, P. V. Gulgunje, Y. Liu, K. K. Gupta, M. G. Kamath, K. M. Lyons, S. Ghoshal, C. Pramanik, L. Giannuzzi, K. Şahin, I. Chasiotis and S. Kumar: "High strength and high modulus carbon fibers." *Carbon* Volume 93, 81–87, **2015**

[10] S. Pimenta and S. T. Pinho: "Recycling carbon fibre reinforced polymers for structural applications: technology review and market outlook." *Waste management (New York, N.Y.)* Volume 31, 378–392, **2011**

[11] C. Olsson, B. Hagström, E. Sjöholm and A. Reimann: "Carbon fibres from lignin-cellulose precursor." *Proc. 18th International Symposium on Wood Fibre and Pulping Chemistry (ISWFPC), Vienna* Volume 2, 126–129, **2015**

[12] Y. Liu, H. G. Chae, Y. H. Choi and S. Kumar: "Preparation of low density hollow carbon fibers by bi-component gel-spinning method." *Journal of Materials Science* Volume 50, 3614–3621, **2015**

[13] P. T. Curtis and S. Travis: "Hollow carbon fibres for high performance polymer composites." *Plastics, Rubber and Composites* Volume 28, 201–209, **1999**

[14] P. V. Gulgunje, B. A. Newcomb, K. Gupta, H. G. Chae, T. K. Tsotsis and S. Kumar: "Low-density and high-modulus carbon fibers from polyacrylonitrile with honeycomb structure." *Carbon* Volume 95, 710–714, **2015**

[15] C. Y. Wang, M. W. Li, Y. L. Wu and C. T. Guo: "Preparation and microstructure of hollow mesophase pitch-based carbon fibers." *Carbon* Volume 36, 1749–1754, **1998**

[16] M. A. Hunt, T. Saito, R. H. Brown, A. S. Kumbhar and A. K. Naskar: "Patterned functional carbon fibers from polyethylene." *Advanced materials (Deerfield Beach, Fla.)* Volume 24, 2386–2389, **2012**

[17] E. Fitzer and L. M. Manocha: *Carbon reinforcements and carbon/carbon composites*. Springer, Berlin, **1998**

[18] B. T. Kelly: *Physics of graphite*. RES mechanica monographs. Applied Science Publ, London, **1981**

[19] H. Jäger, C. Cherif, M. Kirsten, T. Behnisch, D. S. Wolz, R. Böhm and M. Gude: "Influence of processing parameters on the properties of car bon fibres - an overview." *Materialwissenschaft und Werkstofftechnik* Volume 47, 1044–1057, **2016**

[20] M. Guigon, A. Oberlin and G. Desarmot: "Microtexture and structure of some high tensile strength, PAN-based carbon fibres." *Fibre Science and Technology* Volume 20, 55–72, **1984**

[21] M. Guigon, A. Oberlin and G. Desarmot: "Microtexture and structure of some high-modulus, PAN-base carbon fibres." *Fibre Science and Technology* Volume 20, 177–198, **1984**

[22] M. Guigon and A. Oberlin: "Heat-treatment of high tensile strength PAN-based carbon fibres: Microtexture, structure and mechanical properties." *Composites Science and Technology* Volume 27, 1–23, **1986**

[23] D. J. Johnson: "Recent advances in studies of carbon fibre structure." *Philosophical transactions. Series A, Mathematical, physical, and engineering sciences* Volume 294, 443–449, **1980**

[24] D. Chung: *Carbon Fiber Composites*. Elsevier Science, Burlington, **1994**

[25] K. S. Krishnan and N. Ganguli: "Large anisotropy of the electrical conductivity of graphite." *Nature* Volume 144, 667, **1939**

[26] P. R. Wallace: “The band theory of graphite.” *Physical Review* Volume 71, 622, **1947**

[27] M. Owen: “Magnetochemische Untersuchungen. Die thermomagnetischen Eigenschaften der Elemente.” *Annalen der Physik* Volume 342, 657–699, **1912**

[28] K. S. Krishnan: “Magnetic anisotropy of graphite.” *Nature* Volume 133, 174–175, **1934**

[29] R. Sengupta, M. Bhattacharya, S. Bandyopadhyay and A. K. Bhowmick: “A review on the mechanical and electrical properties of graphite and modified graphite reinforced polymer composites.” *Progress in Polymer Science* Volume 36, 638–670, **2011**

[30] N. Oya and D. J. Johnson: “Longitudinal compressive behaviour and microstructure of PAN-based carbon fibres.” *Carbon* Volume 39, 635–645, **2001**

[31] R. J. Diefendorf and E. Tokarsky: “High-performance carbon fibers.” *Polymer Engineering and Science* Volume 15, 150–159, **1975**

[32] S. C. Bennett and D. Johnson: “Electron-microscope studies of structural heterogeneity in PAN-based carbon fibres.” *Carbon* Volume 17, 25–39, **1979**

[33] B. Jones and R. Duncan: “The effect of fibre diameter on the mechanical properties of graphite fibres manufactured from polyacrylonitrile and rayon.” *Journal of Materials Science* Volume 6, 289–293, **1971**

[34] Y. Huang and R. J. Young: “Effect of fibre microstructure upon the modulus of PAN- and pitch-based carbon fibres.” *Carbon* Volume 33, 97–107, **1995**

[35] G. Zhou, Y. Liu, L. He, Q. Guo and H. Ye: "Microstructure difference between core and skin of t700 carbon fibers in heat-treated carbon/carbon composites." *Carbon* Volume 49, 2883–2892, **2011**

[36] G. Zhou, J.-H. Byun, S.-B. Lee, J.-W. Yi, W. Lee, S.-K. Lee, B.-S. Kim, J.-K. Park, S.-G. Lee and L. He: "Nano structural analysis on stiffening phenomena of PAN-based carbon fibers during tensile deformation." *Carbon* Volume 76, 232–239, **2014**

[37] X. Guo, Y. Cheng, Z. Fan, Z. Feng, L. He, R. Liu and J. Xu: "New insights into orientation distribution of high strength polyacrylonitrile-based carbon fibers with skin-core structure." *Carbon* Volume 109, 444–452, **2016**

[38] S. C. Bennett: *Strength-structure relationships in carbon fibres: PhD Thesis*. University of Leeds, **1976**

[39] S. Bennett and D. Johnson: "Fifth London International Carbon and Graphite Conference." *Society of Chemical Industry, London* Volume 377, **1978**

[40] Y. Wang and T. H. Hahn: "AFM characterization of the interfacial properties of carbon fiber reinforced polymer composites subjected to hygrothermal treatments." *Composites Science and Technology* Volume 67, 92–101, **2007**

[41] X. Liu, C. Zhu, J. Guo, Q. Liu, H. Dong, Y. Gu, R. Liu, N. Zhao, Z. Zhang and J. Xu: "Nanoscale dynamic mechanical imaging of the skin–core difference: from PAN precursors to carbon fibers." *Materials Letters* Volume 128, 417–420, **2014**

[42] T. Matsumoto and I. Mochida: "Oxygen distribution in oxidatively stabilized mesophase pitch fiber." *Carbon* Volume 31, 143–147, **1993**

[43] M.-Y. Lv, H.-Y. Ge and J. Chen: "Study on the chemical structure and skin-core structure of polyacrylonitrile-based fibers during stabilization." *Journal of Polymer Research* Volume 16, 513–517, **2009**

[44] C. Blanco, S. Lu, S. Appleyard and B. Rand: "The stabilisation of carbon fibres studied by micro-thermal analysis." *Carbon* Volume 41, 165–171, **2003**

[45] S. Nunna, C. Creighton, N. Hameed, M. Naebe, L. C. Henderson, M. Setty and B. L. Fox: "Radial structure and property relationship in the thermal stabilization of PAN precursor fibres." *Polymer Testing* Volume 59, 203–211, **2017**

[46] E. A. Morris, M. C. Weisenberger, M. G. Abdallah, F. Vautard, H. Grappe, S. Ozcan, F. L. Paulauskas, C. Eberle, D. Jackson, S. J. Mecham and A. K. Naskar: "High performance carbon fibers from very high molecular weight polyacrylonitrile precursors." *Carbon* Volume 101, 245–252, **2016**

[47] W. Watt and W. Johnson: "Mechanism of oxidisation of polyacrylonitrile fibres." *Nature* Volume 257, 210–212, **1975**

[48] F. R. Barnet and M. K. Norr: "A three-dimensional structural model for a high modulus PAN-based carbon fibre." *Composites* Volume 7, 93–99, **1976**

[49] D. Loidl, O. Paris, M. Burghammer, C. Riekel and H. Peterlik: "Direct observation of nanocrystallite buckling in carbon fibers under bending load." *Physical review letters* Volume 95, **2005**

[50] O. Paris, D. Loidl and H. Peterlik: "Texture of PAN- and pitch-based carbon fibers." *Carbon* Volume 40, 551–555, **2002**

[51] D. Loidl, O. Paris, H. Rennhofer, M. Müller and H. Peterlik: "Skin-core structure and bimodal weibull distribution of the strength of carbon fibers." *Carbon* Volume 45, 2801–2805, **2007**

[52] D. Loidl, H. Peterlik, M. Müller, C. Riekel and O. Paris: "Elastic moduli of nanocrystallites in carbon fibers measured by in-situ X-ray microbeam diffraction." *Carbon* Volume 41, 563–570, **2003**

[53] S. Akita, H. Nishijima, Y. Nakayama, F. Tokumasu and K. Takeyasu: "Carbon nanotube tips for a scanning probe microscope: their fabrication and properties." *Journal of Physics D: Applied Physics* Volume 32, 1044–1048, **1999**

[54] M. Treacy, A. Krishnan and P. N. Yianilos: "Inferring physical parameters from images of vibrating carbon nanotubes." *Microscopy and Microanalysis* Volume 6, 317–323, **2000**

[55] E. Fitzer and M. Heine: "Carbon fibre manufacture and surface treatment." In A. R. Bunsell, editor, "Fibre reinforcements for composite materials," Composite Materials Series, 73–148. Elsevier, Amsterdam, **1988**

[56] E. Fitzer, W. Frohs and M. Heine: "Optimization of stabilization and carbonization treatment of PAN fibres and structural characterization of the resulting carbon fibres." *Carbon* Volume 24, 387–395, **1986**

[57] C. Freudenberg: "Textile Faserstoffe." In C. Cherif, editor, "Textile Werkstoffe für den Leichtbau," 39–109. Springer, Berlin, Heidelberg, **2011**

[58] E. Fitzer, M. Heine and G. Jacobsen: "Kohlenstoffasern (Carbonfasern)." In Herbert Heissler, editor, "Verstärkte Kunststoffe in der Luft- und Raumfahrt," 95–128. Verlag W. Kohlhammer, Stuttgart, Berlin, Köln, Mainz, **1986**

[59] S. Chand: "Review carbon fibers for composites." *Journal of Materials Science* Volume 35, 1303–1313, **2000**

[60] P. J. Goodhew, A. J. Clarke and J. E. Bailey: "A review of the fabrication and properties of carbon fibres." *Materials Science and Engineering* Volume 17, 3–30, **1975**

[61] M. Nič, J. Jirát, B. Košata, A. Jenkins and A. McNaught: *IUPAC Compendium of Chemical Terminology*. IUPAC, Research Triagle Park, NC, **2009**

[62] O. P. Bahl, Z. Shen, J. G. Lavin and R. Ross: "Manufacture of carbon fibers." In E. D. Weil, editor, "Carbon fibers, 2nd edition by J. B. Donnet and R. C. Bansal, Marcel Dekker, New York," 1–83. **1992**

[63] A. K. Gupta, D. K. Paliwal and P. Bajaj: "Acrylic precursors for carbon fibers." *Polymer Reviews* Volume 31, 1–89, **1991**

[64] R. Moreton and W. Watt: "The spinning of polyacrylonitrile fibres in clean room conditions for the production of carbon fibres." *Carbon* Volume 12, 543–554, **1974**

[65] J. W. Johnson and D. J. Thorne: "Effect of internal polymer flaws on strength of carbon fibres prepared from an acrylic precursor." *Carbon* Volume 7, 659–661, **1969**

[66] A. Serkov, G. Budnitskii, M. Radishevskii, V. Medvedev and L. Zlatoustova: "Improving carbon fibre production technology." *Fibre Chemistry* Volume 35, 117–121, **2003**

[67] R. B. Mathur, O. P. Bahl and J. Mittal: "Advances in the development of high-performance carbon fibres from PAN precursor." *Composites Science and Technology* Volume 51, 223–230, **1994**

[68] D. P. Bahl, R. B. Mathur and T. L. Dhami: "Modification of polyacrylo nitrile fibres to make them suitable for conversion into high performance carbon fibres." *Materials Science and Engineering* Volume 73, 105–112, **1985**

[69] J. Chen and I. Harrison: "Modification of polyacrylonitrile (PAN) carbon fiber precursor via post-spinning plasticization and stretching in dimethyl formamide (DMF)." *Carbon* Volume 40, 25–45, **2002**

[70] J. Mittal, R. B. Mathur and O. P. Bahl: "Post spinning modification of PAN fibres – a review." *Carbon* Volume 35, 1713–1721, **1997**

[71] S. S. Lin: "Recent developments of carbon fiber in Japan." *SAMPE Journal (Society of Aerospace Material and Process Engineers)* Volume 28, **1992**

[72] B. A. Newcomb and H. G. Chae: "The properties of carbon fibers." In A. R. Bunsell, editor, "Handbook of Properties of Textile and Technical Fibres," The Textile Institute Book Series, 841–871. Elsevier, **2018**

[73] E. Fitzer and W. Frohs: "The influence of the carbonization temperature and post treatment conditions on the properties of pan-based carbon fibers." *Proceedings of the International Carbon Conference* 298–300, **1988**

[74] R. Moreton, W. Watt and W. Johnson: "Carbon fibres of high strength and high breaking strain." *Nature* Volume 213, 690–691, **1967**

[75] K. Badii, G. Golkarnarenji, A. S. Milani, M. Naebe and H. Khayyam: "A comprehensive chemical model for the preliminary steps of the thermal stabilization process in a carbon fibre manufacturing line." *Reaction Chemistry & Engineering* Volume 3, 959–971, **2018**

[76] Z. Bashir: "A critical review of the stabilisation of polyacrylonitrile." *Carbon* Volume 29, 1081–1090, **1991**

[77] S. Ozbek and D. H. Isaac: "Carbon fiber processing: effects of hot stretching on mechanical properties." *Materials and Manufacturing Processes* Volume 9, 199–219, **1994**

[78] D. H. Isaac, S. Ozbek and J. G. Francis: "Processing of carbon fibers: texture enhancement induced by hot stretching." *Materials and Manufacturing Processes* Volume 9, 179–197, **1994**

[79] S. Ozbek and D. H. Isaac: "Strain-induced density changes in pan-based carbon fibres." *Carbon* Volume 38, 2007–2016, **2000**

[80] H. Schürmann: *Konstruieren mit Faser-Kunststoff-Verbunden*. Springer, Berlin, Heidelberg, **2007**

[81] E. Fitzer and R. Weiss: "Effect of surface treatment and sizing of C-fibres on the mechanical properties of CFR thermosetting and thermoplastic polymers." *Carbon* Volume 25, 455–467, **1987**

[82] F. R. Jones: "Interphase formation and control in fibre composite materials." *Key Engineering Materials* Volume 116-117, 41–60, **1995**

[83] S.-R. Shamsuddin, K. K. Chee Ho, K.-Y. Lee, J. M. Hodgkinson and A. Bismarck: In L. Nicolais, A. Borzacchielo and S. M. Lee, editors, "Wiley encyclopedia of composites," 1–17. Wiley, Hoboken, NJ, **2012**

[84] S. Pimenta and S. T. Pinho: "The effect of recycling on the mechanical response of carbon fibres and their composites." *Composite Structures* Volume 94, 3669–3684, **2012**

[85] G. Osumi and E. Kato: "A review of testing instruments and an improved recording fibre tester." *Journal of the Textile Institute Transactions* Volume 28, T129–T144, **1937**

[86] C. O'Neill: "Experiments and observations upon cotton." *Manchester Literary and Philo-sophical Society Memoirs* **1863**

[87] E. H. Pressley: "A cotton fiber strength tester." *American Society for Testing and Materials Bulletin* Volume 118, 13–17, **1942**

[88] *ASTM standard D1445/D1445M – 12: Standard Test Method for Breaking Strength and Elongation of Cotton Fibers (Flat Bundle Method).* West Conshohocken, PA, **2012**

[89] S. Williams and E. V. Painter: "A study of the pressley cotton fiber strength tester." *Textile Research* Volume 15, 403–412, **1945**

[90] unknown author: "Cotton fiber strength study." *Textile Research* Volume 13, 17–19, **1943**

[91] A. J. Turner and V. Venkataraman: "The foundation of yarn-strength and yarn-extension: Part V – The prediction of the spinning value of cotton from its fibre-properties." *Journal of the Textile Institute Transactions* Volume 25, 1–48, **1934**

[92] D. D. Fang, editor: *Cotton Fiber: Physics, Chemistry and Biology.* Springer International Publishing, Cham, **2018**

[93] A. S. Watson and R. L. Smith: "An examination of statistical theories for fibrous materials in the light of experimental data." *Journal of Materials Science* Volume 20, 3260–3270, **1985**

[94] R. L. Smith, S. L. Phoenix, M. R. Greenfield, R. B. Henstenburg and R. E. Pitt: "Lower-tail approximations for the probability of failure of three- dimensional fibrous composites with hexagonal geometry." *Proceedings of the Royal Society A: Mathematical, Physical and Engineering Sciences* Volume 388, 353–391, **1983**

[95] A. Cowking, A. Attou, A. M. Siddiqui, M. A. S. Sweet and R. Hill: "Testing E-glass fibre bundles using acoustic emission." *Journal of Materials Science* Volume 26, 1301–1310, **1991**

[96] P. K. Mallick: *Fiber-Reinforced Composites: Materials, Manufacturing, and Design, Third Edition.* Mechanical Engineering. CRC Press, **2007**

[97] R. S. Orr, L. C. Weiss and J. N. Grant: "The relation of single-fiber to flat-bundle strength and elongation of cotton." *Textile Research Journal* Volume 25, 939–946, **1955**

[98] V. E. Gonsalves: "Determination of Denier and strength of single filaments by vibroscope and Heim tensile tester." *Textile Research Journal* Volume 17, 369–375, **1947**

[99] P. E. Sasser, F. M. Shofner, Y. T. Chu, C. K. Shofner and M. G. Townes: "Interpretations of single fiber, bundle, and yarn tenacity data." *Textile Research Journal* Volume 61, 681–690, **1991**

[100] C. D. Delhom, X. Cui and D. P. Thibodeaux: "Single fiber testing via Favimat." In "2010 Beltwide Cotton Conference. New Orleans, Louisiana," 4–7. **2010**

[101] R. Hill and E. U. Okoroafor: "Weibull statistics of fibre bundle failure using mechanical and acoustic emission testing: the influence of interfibre friction." *Composites* Volume 26, 699–705, **1995**

[102] Hughes, JDH and Morley, H and Jackson, EE: "Aligned carbon fibre composite which approaches theoretical strength." *Journal of Physics D: Applied Physics* Volume 13, 921–936, **1980**

[103] B. D. Coleman: "On the strength of classical fibres and fibre bundles." *Journal of the Mechanics and Physics of Solids* Volume 7, 60–70, **1958**

[104] J. J. Hebert, D. P. Thibodeaux, F. M. Shofner, J. K. Singletary and D. B. Patelke: "A new single fiber tensile tester." *Textile Research Journal* Volume 65, 440–444, **1995**

[105] J. A. Foulk and D. D. McAlister: "Single cotton fiber properties of low, ideal, and high micronaire values." *Textile Research Journal* Volume 72, 885–891, **2002**

[106] M. G. Huson, J. S. Church, A. A. Kafi, A. L. Woodhead, J. Khoo, M. Kiran, J. E. Bradby and B. L. Fox: "Heterogeneity of carbon fibre." *Carbon* Volume 68, 240–249, **2014**

[107] U. Mörschel, C. Faymonville and E. Ingelsberger: "Automatic linear-density and tensile testing on carbon fibers." *Chem. Fibers Int* Volume 4, 248–249, **2008**

[108] A. Fraisse, J. Beauson, P. Brøndsted and B. Madsen: "Thermal recycling and re-manufacturing of glass fibre thermosetting composites." *IOP Conference Series: Materials Science and Engineering* Volume 139, 1–10, **2016**

[109] J. Huether, P. Rupp, I. Kohlschreiber and K. A. Weidenmann: "An enhanced method to determine the young's modulus of technical single fibres by means of high resolution digital image correlation." *Measurement Science and Technology* Volume 29, 1–13, **2018**

[110] J. Huether, D. Esse, P. Rupp, M. Seitz and K. A. Weidenmann: "Enhanced strain and stiffness evaluation of virgin and recycled single carbon fibres by means of one-dimensional digital image correlation." *Measurement Science and Technology* Volume 30, 1–12, **2019**

[111] British Standards: "ISO 11566:1996: Carbon fibre - Determination of the tensile properties of single-filament specimens: Carbon materials.", **1996**

[112] C28 Committee: *ASTM standard D 3379–75: Test Method for Tensile Strength and Youngs Modulus for High-Modulus Single-Filament Materials: Standard Test Method for Tensile Strength and Young's Modulus for High-Modulus Single-Filament Materials (Withdrawn 1998)*. ASTM International, West Conshohocken, PA, **1975**

[113] C28 Committee: *ASTM standard C1557–14: Test Method for Tensile Strength and Young's Modulus of Fibers*. ASTM International, West Conshohocken, PA, **2014**

[114] K. G. Dassios, M. Steen and C. Filiou: "Mechanical properties of alumina Nextel 720 fibres at room and elevated temperatures: tensile bundle testing." *Materials Science and Engineering: A* Volume 349, 63–72, **2003**

[115] C.-T. Li and N. R. Langley: "Improvement in fiber testing of high-modulus single-filament materials." *Journal of the American Ceramic Society* Volume 68, 202–204, **1985**

[116] M. Kant and D. Penumadu: "Dynamic mechanical characterization for nonlinear behavior of single carbon fibers." *Composites Part A: Applied Science and Manufacturing* Volume 66, 201–208, **2014**

[117] K. Hendrickx, D. Depuydt, A. W. van Vuure, A. Willem and J. Ivens: "The relationship between the tensile properties of natural fibres and their ud composites." *Proceedings of ECCM 17 - 17th European conference on composite materials* 1 – 8, **2016**

[118] J. R. Davis: *Tensile testing*. ASM International, Materials Park, Ohio, 2nd ed. edition, **2004**

[119] M. Sutton, W. Wolters, W. Peters, W. Ranson and S. McNeill: "Determination of displacements using an improved digital correlation method." *Image and vision computing* Volume 1, 133–139, **1983**

[120] B. Pan, K. Qian, H. Xie and A. Asundi: "Two-dimensional digital image correlation for in-plane displacement and strain measurement: a review." *Measurement Science and Technology* Volume 20, 1–17, **2009**

[121] M. Grediac: "The use of full-field measurement methods in composite material characterization: interest and limitations." *Composites Part A: Applied Science and Manufacturing* Volume 35, 751–761, **2004**

[122] J. Xavier, A. Sousa, J. J. L. Morais, V. M. J. Filipe and M. Vaz: "Measuring displacement fields by cross-correlation and a differential technique: experimental validation." *Optical Engineering* Volume 51, 1–12, **2012**

[123] Friebe H, Klein, M: "Optische Bewegungs- und Verformungsanalyse auf Basis applizierter Muster und Marken." In "5. VDI Fachtagung optische Messung von Funktionsflächen 2016," 47–60. **2016**

[124] D. Winter: *Optische Verschiebungsmessung nach dem Objektrasterprinzip: PhD Thesis.* Technische Universität Braunschweig, **1993**

[125] C. Reder, D. Loidl, S. Puchegger, D. Gitschthaler, H. Peterlik, K. Kromp, G. Khatibi, A. Betzwar-Kotas, P. Zimprich and B. Weiss: "Non-contacting strain measurements of ceramic and carbon single fibres by using the laser-speckle method." *Composites Part A: Applied Science and Manufacturing* Volume 34, 1029–1033, **2003**

[126] M. Anwander, B. G. Zagar, B. Weiss and H. Weiss: "Noncontacting strain measurements at high temperatures by the digital laser speckle technique." *Experimental Mechanics* Volume 40, 98–105, **2000**

[127] D. Depuydt, K. Hendrickx, W. Biesmans, J. Ivens and A. W. van Vuure: "Digital image correlation as a strain measurement technique for fibre tensile tests." *Composites Part A: Applied Science and Manufacturing* Volume 99, 76–83, **2017**

[128] D. Coimbra, R. Greenwood and K. Kendall: "Tensile testing of ceramic fibres by video extensometry." *Journal of Materials Science* Volume 35, 3341–3345, **2000**

[129] R.-B. Adusumalli, U. Müller, H. Weber, T. Roeder, H. Sixta and W. Gindl: "Tensile testing of single regenerated cellulose fibres." In "Macromolecular Symposia," Volume 244, 83–88. **2006**

[130] I. Burgert, K. Frühmann, J. Keckes, P. Fratzl and S. E. Stanzl-Tschegg: "Microtensile testing of wood fibers combined with video extensometry for efficient strain detection." *Holzforschung* Volume 57, 661–664, **2003**

[131] I. Bronstein, K. Semendjajew, G. Musiol and H. Mühlig: *Taschenbuch der Mathematik*. Verlag Harri Deutsch, Frankfurt am Main, **2008**

[132] T. Bonnesen and W. Fenchel: *Theorie der konvexen Körper*. Springer, Berlin, Heidelberg, **1934**

[133] R. Schneider: *Convex bodies: The Brunn-Minkowski theory*, Volume 44 of *Encyclopedia of mathematics and its applications*. Cambridge University Press, Cambridge and New York, **1993**

[134] P. E. McMahon: "Graphite fiber tensile property evaluation." In J. M. Whitney, editor, "Analysis of the Test Methods for High Modulus Fibers and Composites," 367–367–23. ASTM International, West Conshohocken, PA, **1973**

[135] J. V. Nabais, P. Carrott and M. Ribeiro Carrott: "From commercial textile fibres to activated carbon fibres: chemical transformations." *Materials Chemistry and Physics* Volume 93, 100–108, **2005**

[136] K. Naito, Y. Tanaka, J.-M. Yang and Y. Kagawa: "Tensile properties of ultrahigh strength PAN-based, ultrahigh modulus pitch-based and high ductility pitch-based carbon fibers." *Carbon* Volume 46, 189–195, **2008**

[137] K. Naito, J.-M. Yang, Y. Tanaka and Y. Kagawa: "The effect of gauge length on tensile strength and Weibull modulus of polyacrylonitrile (PAN)- and pitch-based carbon fibers." *Journal of Materials Science* Volume 47, 632–642, **2012**

[138] K. Naito, Y. Tanaka and J.-M. Yang: "Transverse compressive properties of polyacrylonitrile (PAN)-based and pitch-based single carbon fibers." *Carbon* Volume 118, 168–183, **2017**

[139] S. Kumar, D. P. Anderson and A. S. Crasto: "Carbon fibre compressive strength and its dependence on structure and morphology" *Journal of Materials Science* Volume 28, 423–439, **1993**

[140] M. Endo: "Structure of mesophase pitch-based carbon fibres." *Journal of Materials Science* Volume 23, 598–605, **1988**

[141] J.D.H. Hughes: "Strength and modulus of current carbon fibres." *Carbon* Volume 24, 551–556, **1986**

[142] T. Cho, Y. S. Lee, R. Rao, A. M. Rao, D. D. Edie and A. A. Ogale: "Structure of carbon fiber obtained from nanotube-reinforced mesophase pitch." *Carbon* Volume 41, 1419–1424, **2003**

[143] S.-S. Tzeng and F.-Y. Chang: "EMI shielding effectiveness of metal-coated carbon fiber-reinforced ABS composites." *Materials Science and Engineering: A* Volume 302, 258–267, **2001**

[144] D. L. Vezie and W. W. Adams: "High resolution scanning electron microscopy of pan-based and pich-based carbon fibre." *Journal of Materials Science Letters* Volume 9, 883–887, **1990**

[145] X. Liu, R. Wang, Z. Wu and W. Liu: "The effect of triangle-shape carbon fiber on the flexural properties of the carbon fiber reinforced plastics." *Materials Letters* Volume 73, 21–23, **2012**

[146] L. P. Kobets and I. S. Deev: "Carbon fibres: structure and mechanical properties." *Composites Science and Technology* Volume 57, 1571–1580, **1998**

[147] W. Xie, H. Cheng, Z. Chu, Z. Chen and C. Long: "Effect of carbonization temperature on the structure and microwave absorbing properties of hollow carbon fibres." *Ceramics International* Volume 37, 1947–1951, **2011**

[148] S.-J. Park, M.-K. Seo, H.-B. Shim and K.-Y. Rhee: "Effect of different cross-section types on mechanical properties of carbon fibers-reinforced cement composites." *Materials Science and Engineering: A* Volume 366, 348–355, **2004**

[149] S.-Y. Eom and S.-K. Ryu: "Properties of differently shaped activated carbon fibers." *Korean Journal of Chemical Engineering* Volume 27, 1592–1595, **2010**

[150] O. Saburow, J. Huether, R. Maertens, A. Trauth, Y. Kechaou, F. Henning and K. A. Weidenmann: "A direct process to reuse dry fiber production waste for recycled carbon fiber bulk molding compounds." *Procedia CIRP* Volume 66, 265–270, **2017**

[151] Z. Li, D. Zhang, W. Qin and D. Geng: "Feasibility study on the rotary ultrasonic elliptical machining for countersinking of carbon fiber–reinforced plastics." *Proceedings of the Institution of Mechanical Engineers, Part B: Journal of Engineering Manufacture* Volume 231, 2347–2358, **2017**

[152] O. Paris, D. Loidl, M. Müller, H. Lichtenegger and H. Peterlik: "Cross-sectional texture of carbon fibres analysed by scanning microbeam X-ray diffraction." *Journal of Applied Crystallography* Volume 34, 473–479, **2001**

[153] G. Daumit and Y. Ko: "A unique approach to carbon fiber precursor development." *High tech: The way into the nineties – Proceedings of the 7th International Conference of the Society for the Advancement of Material and Process Engineering European Chapter* Volume 7, 201–213, **1986**

[154] J. S. Marcuzzo, C. Otani, H. A. Polidoro and S. Otani: "Influence of thermal treatment on porosity formation on carbon fiber from textile PAN." *Materials Research* Volume 16, 137–144, **2013**

[155] Deutsches Institut für Normung: "DIN EN ISO 1973: Bestimmung der Feinheit - Gravimetrisches Verfahren und Schwingungsverfahren." **1995**

[156] I. M. Kowalski: "Characterizing the tensile stress-strain nonlinearity of polyacrylonitrile-based carbon fibers." In J. D. Whitcomb, editor, "Composite materials: testing and design (eight conference), Charleston, South Carolina," ASTM STP, 205–205–12. ASTM, Philadelphia, Pa., **1988**

[157] W. Blohm, H. Sikora and A. Beining: "Durchmessermessung mit Beugungssäumen sowie elektronische Verschmutzungskorrektur - European Patent Office - EP 0924493 B1: Europäische Patentschrift.", **1999**

[158] S. A. Khodier: "Measurement of wire diameter by optical diffraction." *Optics & Laser Technology* Volume 36, 63–67, **2004**

[159] J. Troll and C. Baker: "US Patent 3659950A – Laser apparatus for detecting the size and form of filamentary material by measuring diffracted light.", **1972**

[160] M. E. Fedorov: "Optical laser diffraction transducer for measuring single-wire electric cable eccentricity." *IOP Conf. Ser.: Mater. Sci. Eng.* Volume 81, 1–10, **2015**

[161] H. Wang and R. Valdivia-Hernandez: "Laser scanner and diffraction pattern detection: a novel concept for dynamic gauging of fine wires." *Measurement Science and Technology* Volume 6, 452–457, **1995**

[162] M. Koedam: "Determination of small dimensions by diffraction of a laser beam." *Philips technical review* Volume 27, 208–211, **1966**

[163] A. R. Bunsell, S. Joannès and A. Marcellan: "Testing and characterization of fibers." In A. R. Bunsell, editor, "Handbook of tensile properties of textile and technical fibres," Woodhead publishing in textiles, 21–55. Woodhead Publishing, Oxford, **2009**

[164] E. Bernabeu, I. Serroukh and L. M. Sanchez-Brea: "Geometrical model for wire optical diffraction selected by experimental statistical analysis." *Optical Engineering* Volume 38, 1319–1325, **1999**

[165] G. B. J. de Boer, C. de Weerd, D. Thoenes and H. W. J. Goossens: "Laser diffraction spectrometry: Fraunhofer diffraction versus Mie scattering." *Particle & Particle Systems Characterization* Volume 4, 14–19, **1987**

[166] R. G. Greenler, J. W. Hable and P. O. Slane: "Diffraction around a fine wire: how good is the single–slit approximation?" *American Journal of Physics* Volume 58, 330–331, **1990**

[167] D. Lebrun, S. Belaid, C. Ozkul, K. F. Ren and G. Grehan: "Enhancement of wire diameter measurements: comparison between Fraunhofer diffraction and Lorenz–Mie theory." *Optical Engineering* Volume 35, 946, **1996**

[168] W. Tang, Y. Zhou and J. Zhang: "Improvement on theoretical model for thin-wire and slot measurement by optical diffraction." *Measurement Science and Technology* Volume 10, 119–123, **1999**

[169] J. Xie, Y. Qiu, H. Ming and C. Li: "Light polarization effect in measurement of thin wire diameter by laser diffraction and its explanation with boundary diffraction wave." *Journal of Applied Physics* Volume 69, 6899–6903, **1991**

[170] J. C. Martínez-Anton, I. Serroukh and E. Bernabeu: "On Babinet's principle and a diffraction-interferometric technique to determine the diameter of cylindrical wires." *Metrologia* Volume 38, 125–134, **2001**

[171] I. Krucinska and T. Stypka: "Direct measurement of the axial poisson's ratio of single carbon fibres." *Composites Science and Technology* Volume 41, 1–12, **1991**

[172] C.-T. Li and J. V. Tietz: "Improved accuracy of the laser diffraction technique for diameter measurement of small fibres." *Journal of Materials Science* Volume 25, 4694–4698, **1990**

[173] S. Meretz, T. Linke, E. Schulz, A. Hampe and M. Hentschel: "Diameter measurement of small fibres: laser diffraction and scanning electron microscopy technique results do not differ systematically." *Journal of Materials Science Letters* Volume 11, 1471–1472, **1992**

[174] A. Bismarck, M. Pfaffernoschke, J. Springer and E. Schulz: "Polystyrene-grafted carbon fibers: surface properties and adhesion to polystyrene." *Journal of Thermoplastic Composite Materials* Volume 18, 307–331, **2005**

[175] A. Bismarck, A. F. Lee, A. S. Saraç, E. Schulz and K. Wilson: "Electrocoating of carbon fibres: a route for interface control in carbon fibre reinforced poly methylmethacrylate?" *Composites Science and Technology* Volume 65, 1564–1573, **2005**

[176] A. Perry, B. Ineichen and B. Eliasson: "Fibre diameter measurement by laser diffraction." *Journal of Materials Science* Volume 9, 1376–1378, **1974**

[177] A. Perry, K. Phillips and E. De Lamotte: "The mechanical properties of carbon fibres." *Fibre Science and Technology* Volume 3, 317–319, **1971**

[178] A. Madroñero and C. Merino: "Some geometrical singularities in the characterization of vapor grown carbon fibers using laser diffraction technique." *Materials Research Bulletin* Volume 33, 1503–1515, **1998**

[179] S. J. Pickering, T. A. Turner, F. Meng, C. N. Morris, J. P. Heil, K. H. Wong and S. Melendi-Espina: "Developments in the fluidised bed process for fibre recovery from thermoset composites." In "2nd Annual Composites and Advanced Materials Expo, CAMX 2015; Dallas Convention CenterDallas; United States," 2384–2394. **2015**

[180] H. Fischer and H. Schmid: "Qualitätskontrolle für rezyklierte Carbonfasern." *Kunststoffe* Volume 103, 88–91, **2013**

[181] H. Kleinhans and L. Salmén: "Development of lignin carbon fibers: evaluation of the carbonization process." *Journal of Applied Polymer Science* Volume 133, 119, **2016**

[182] A. P. Nowak, J. Hagberg, S. Leijonmarck, H. Schweinebarth, D. Baker, A. Uhlin, P. Tomani and G. Lindbergh: "Lignin-based carbon fibers for renewable and multifunctional lithium-ion battery electrodes." *Holzforschung* Volume 72, 81–90, **2018**

[183] A. Bengtsson, J. Bengtsson, C. Olsson, M. Sedin, K. Jedvert, H. Theliander and E. Sjöholm: "Improved yield of carbon fibres from cellulose and kraft lignin." *Holzforschung* Volume 72, 1007–1016, **2018**

[184] A. Awal, G. Cescutti, S. B. Ghosh and J. Müssig: "Interfacial studies of natural fibre/polypropylene composites using single fibre fragmentation test (SFFT)." *Composites Part A: Applied Science and Manufacturing* Volume 42, 50–56, **2011**

[185] J. Müssig: "Cotton fibre-reinforced thermosets versus Ramie composites: a comparative study using petrochemical- and agro-based resins." *Journal of Polymers and the Environment* Volume 16, 94–102, **2008**

[186] N. Graupner: "Application of lignin as natural adhesion promoter in cotton fibre-reinforced poly(lactic acid) (PLA) composites." *Journal of Materials Science* Volume 43, 5222–5229, **2008**

[187] N. Graupner, A. S. Herrmann and J. Müssig: "Natural and man-made cellulose fibre-reinforced poly (lactic acid)(pla) composites: an overview about mechanical characteristics and application areas." *Composites Part A: Applied Science and Manufacturing* Volume 40, 810–821, **2009**

[188] N. Graupner: "Improvement of the mechanical properties of biodegradable hemp fiber reinforced poly(lactic acid) (pla) composites by the admixture of man-made cellulose fibers." *Journal of Composite Materials* Volume 43, 689–702, **2009**

[189] H. Fischer, E. Werwein and N. Graupner: "Nettle fibre (urtica dioica l.) reinforced poly(lactic acid): a first approach." *Journal of Composite Materials* Volume 46, 3077–3087, **2012**

[190] J. L. Thomason, J. Carruthers, J. Kelly and G. Johnson: "Fibre cross-section determination and variability in sisal and flax and its effects on fibre performance characterisation." *Composites Science and Technology* Volume 71, 1008–1015, **2011**

[191] T. Morimoto and T. Ogasawara: "Potential strength of Nicalon, Hi Nicalon, and Hi Nicalon Type S monofilaments of variable diameters." *Composites Part A: Applied Science and Manufacturing* Volume 37, 405–412, **2006**

[192] D. J. Montgomery and W. T. Milloway: "The vibroscopic method for determination of fiber cross-sectional area." *Textile Research Journal* Volume 22, 729–735, **1952**

[193] K. M. Lyons, B. A. Newcomb, K. J. McDonald, H. G. Chae and S. Kumar: "Development of single filament testing procedure for polyacrylonitrile precursor and polyacrylonitrile-based carbon fibers." *Journal of Composite Materials* Volume 49, 2231–2240, **2015**

[194] K. M. Beggs, L. Servinis, T. R. Gengenbach, M. G. Huson, B. L. Fox and L. C. Henderson: "A systematic study of carbon fibre surface grafting via in situ diazonium generation for improved interfacial shear strength in epoxy matrix composites." *Composites Science and Technology* Volume 118, 31–38, **2015**

[195] F. Gnädinger, P. Middendorf and B. Fox: "Interfacial shear strength studies of experimental carbon fibres, novel thermosetting polyurethane and epoxy matrices and bespoke sizing agents." *Composites Science and Technology* Volume 133, 104–110, **2016**

[196] H. U. Sokoli, J. Beauson, M. E. Simonsen, A. Fraisse, P. Brøndsted and E. G. Søgaard: "Optimized process for recovery of glass- and carbon fibers with retained mechanical properties by means of near- and supercritical fluids." *The Journal of Supercritical Fluids* Volume 124, 80–89, **2017**

[197] I. R. Titze and E. J. Hunter: "Normal vibration frequencies of the vocal ligament." *The Journal of the Acoustical Society of America* Volume 115, 2264–2269, **2004**

[198] I. M. Djordjević, D. R. Sekulić, M. N. Mitrić and M. M. Stevanović: "Non-hookean elastic behavior and crystallite orientation in carbon fibers." *Journal of Composite Materials* Volume 44, 1717–1727, **2010**

[199] C. P. Beetz and G. W. Budd: "Strain modulation measurements of stiffening effects in carbon fibers." *Review of Scientific Instruments* Volume 54, 1222–1226, **1983**

[200] Y. Huang and R. J. Young: "Non-linear elasticity in carbon fibres." *Journal of Materials Science Letters* Volume 12, 92–95, **1993**

[201] P. Arsenovic, H. Jiang, R. K. Eby, W. W. Adams and J. M. Liu: "Nonlinear elasticity of carbon fibers." In B. MacEnaney, editor, "Carbon '88," IOP Publ, Bristol, **1988**

[202] P. Arsenovic, H. Jiang, R. K. Eby and J. M. Liu: "Effects of applied stress and temperature on the nonlinear elastic properties of graphite fibers." *Naval Surface Warfare Center* Volume 1, 1–32, **1987**

[203] M. Shioya and A. Takaku: "Rotation and extension of crystallites in carbon fibers by tensile stress." *Carbon* Volume 32, 615–619, **1994**

[204] M. Shioya, E. Hayakawa and A. Takaku: "Non-hookean stress-strain response and changes in crystallite orientation of carbon fibres." *Journal of Materials Science* Volume 31, 4521–4532, **1996**

[205] A. E. Bogdanovich, M. Karahan, S. V. Lomov and I. Verpoest: "Quasi-static tensile behavior and damage of carbon/epoxy composite reinforced with 3D non-crimp orthogonal woven fabric." *Mechanics of Materials* Volume 62, 14–31, **2013**

[206] J. Harper and T. Heumann: "The strain dependence of elastic modulus in unidirectional composites." *Proceedings of the Second International Conference on Testing, Evaluation and Quality Control of Composites–TEQC 87* 189–193, **1987**

[207] T. Ishikawa, M. Matsushima and Y. Hayashi: "Hardening non-linear behaviour in longitudinal tension of unidirectional carbon composites." *Journal of Materials Science* Volume 20, 4075–4083, **1985**

[208] W. H. Van Dreumel and J. L. Kamp: "Non hookean behaviour in the fibre direction of carbon-fibre composites and the influence of fibre waviness on the tensile properties." *Journal of Composite Materials* Volume 11, 461–469, **1977**

[209] P. A. Lagace: "Nonlinear stress-strain behavior of graphite/epoxy laminates." *AIAA Journal* Volume 23, 1583–1589, **1985**

[210] I. Djordjević, D. Sekulić and M. Stevanovic: "Non-linear elastic behavior of carbon fibres of different structural and mechanical characteristic." *Journal of the Serbian Chemical Society* Volume 72, 513–521, **2007**

[211] N. Toyama and J. Takatsubo: "An investigation of non-linear elastic behavior of CFRP laminates and strain measurement using Lamb waves." *Composites Science and Technology* Volume 64, 2509–2516, **2004**

[212] T. B. Stecenko and M. M. Stevanovi: "Variation of elastic moduli with strain in carbon/epoxy laminates." *Journal of Composite Materials* Volume 24, 1152–1158, **1990**

[213] A. A. Griffith: "The phenomena of rupture and flow in solids." *Phil, Trans, Roy. Soc. A221* Volume 221, 163–198, **1921**

[214] W. Whitney and R. Kimmel: "Griffith equation and carbon fibre strength." *Nature Physical Science* Volume 237, 93–94, **1972**

[215] S. C. Bennett, D. J. Johnson and W. Johnson: "Strength-structure relationships in PAN-based carbon fibres." *Journal of Materials Science* Volume 18, 3337–3347, **1983**

[216] W. Reynolds and J. Sharp: "Crystal shear limit to carbon fibre strength." *Carbon* Volume 12, 103–110, **1974**

[217] W. Weibull: "A statistical distribution function of wide applicability." *Journal of applied mechanics* Volume 103, 293–297, **1951**

[218] W. Padgett, S. Durham and A. Mason: "Weibull analysis of the strength of carbon fibers using linear and power law models for the length effect." Volume 29, 1873–1884, **1995**

[219] C. P. Beetz: "The analysis of carbon fibre strength distributions exhibiting multiple modes of failure." *Fibre Science and Technology* Volume 16, 45–59, **1982**

[220] J. L. Thomason: "On the application of Weibull analysis to experimentally determined single fibre strength distributions." *Composites Science and Technology* Volume 77, 74–80, **2013**

[221] Z. Chi, T.-W. Chou and G. Shen: "Determination of single fibre strength distribution from fibre bundle testings." *Journal of Materials Science* Volume 19, 3319–3324, **1984**

[222] F. W. Zok: "On weakest link theory and Weibull statistics." *Journal of the American Ceramic Society* Volume 100, 1265–1268, **2017**

[223] L. Peebles, Y. Yanovsky, A. Sirota, V. Bogdanov and P. Levit: "Mechanical properties of carbon fibers." In E. D. Weil, editor, "Carbon fibers, 2nd edition by J. B. Donnet and R. C. Bansal, Marcel Dekker, New York," 311–369. **1992**

[224] E. De Lamotte and A. Perry: "Diameter and strain-rate dependence of the ultimate tensile strength and Young's modulus of carbon fibres." *Fibre Science and Technology* Volume 3, 157–166, **1970**

[225] D. D. Edie and R. J. Diefendorf: "Carbon fiber manufacturing." In J. D. Buckley and D. D. Edie, editors, "Carbon-carbon materials and composites," 19–39. Noyes Publications, Park Ridge, N.J., U.S.A, **2010**

[226] M. Flemming and S. Roth: *Faserverbundbauweisen Eigenschaften.* Springer Berlin Heidelberg, Berlin, Heidelberg, **2003**

[227] J. N. Grant and O. W. Morlier: "Relation of specific strength of cotton fibers to fiber length and testing method." *Textile Research Journal* Volume 18, 481–487, **1948**

[228] N. Pan, H. Chen, J. Thompson, M. Inglesby, S. Khatua, X. Zhang and S. Zeronian: "The size effects on the mechanical behaviour of fibres." *Journal of Materials Science* Volume 32, 2677–2685, **1997**

[229] S. Chwastiak, J. B. Barr and R. Didchenko: "High strength carbon fibers from mesophase pitch." *Carbon* Volume 17, 49–53, **1979**

[230] J. Watanabe, F. Tanaka, H. Okuda and T. Okabe: "Tensile strength distribution of carbon fibers at short gauge lengths." *Advanced Composite Materials* Volume 23, 535–550, **2014**

[231] S. Deng, L. Ye, Y.-W. Mai and H.-Y. Liu: "Evaluation of fibre tensile strength and fibre/matrix adhesion using single fibre fragmentation tests." *Composites Part A: Applied Science and Manufacturing* Volume 29, 423–434, **1998**

[232] R. Moreton: "The effect of gauge length on the tensile strength of R.A.E. carbon fibres." *Fibre Science and Technology* Volume 1, 273–284, **1969**

[233] J. F. Villeneuve, R. Naslain, R. Fourmeaux and J. Sevely: "Longitudinal/radial thermal expansion and poisson ratio of some ceramic fibres as measured by transmission electron microscopy." *Composites Science and Technology* Volume 49, 89–103, **1993**

[234] J. Hughes: "The carbon fibre/epoxy interface – a review." *Composites Science and Technology* Volume 41, 13–45, **1991**

[235] F. Henning and E. Moeller: "Faserverstärkte Kunststoffe." In F. Henning and E. Moeller, editors, "Handbuch Leichtbau," 393–411. Carl Hanser Verlag GmbH & Co. KG, München, **2011**

[236] F. Zwicky: "On the imperfections of crystals." *Proceedings of the National Academy of Sciences of the United States of America* Volume 15, 253–259, **1929**

[237] G. Slayter: "Two-phase materials." *Scientific American* Volume 206, 124–135, **1962**

[238] F. Henning and E. Moeller, editors: *Handbuch Leichtbau.* Carl Hanser Verlag GmbH & Co. KG, München, **2011**

[239] R. Stewart: "Lightweighting the automotive market." *Reinforced Plastics* Volume 53, 14–21, **2009**

[240] E. Witten, V. Mathes, M. Sauer and M. Kühnel: "Composite market report 2018: market developments, trends, outlook and challenges." *Study by AVK and Carbon Composites, available online: https://www.avk-tv.de/files/20181115_avk_ccev_market_report_2018_final.pdf; 12.11.2019* 1–59, **2018**

[241] J. Woidasky: "Recycling von faserverstärkten Verbunden - Weiterentwicklung des Recyclings von faserverstärkten Verbunden." In K. J. Thome-Kozmiensky and D. Goldmann, editors, "Recycling und Rohstoffe," 240–259. TK Vivis, Neuruppin, **2013**

[242] F. Henning and E. Moeller: "Verarbeitung faserverstärkter Kunststoffe." In F. Henning and E. Moeller, editors, "Handbuch Leichtbau," 603–668. Carl Hanser Verlag GmbH & Co. KG, München, **2011**

[243] P. Beardmore: "Composite structures for automobiles." *Composite Structures* Volume 5, 163–176, **1986**

[244] F. E. Devine: "Polyester moulding materials in automotive underbonnet environments." *Composites* Volume 14, 353–358, **1983**

[245] M. F. Ashby: *Materials Selection in Mechanical Design.* Elsevier professional, Oxford, 3rd edition edition, **2005**

[246] C. Cherif: "Textile Prozesskette und Einordnung der textilen Halbzeuge." In C. Cherif, editor, "Textile Werkstoffe für den Leichtbau," 9–37. Springer, Berlin, Heidelberg, **2011**

[247] H. Kong, A. P. Mouritz and R. Paton: "Tensile extension properties and deformation mechanisms of multiaxial non-crimp fabrics." *Composite Structures* Volume 66, 249–259, **2004**

[248] F. Schirmaier: *Experimentelle Untersuchung und Simulation des Umformverhaltens nähgewirkter unidirektionaler Kohlenstofffasergelege: PhD Thesis.* Karlsruher Schriftenreihe Fahrzeugsystemtechnik. KIT Scientific Publishing, Karlsruhe, **2017**

[249] A. Hohmann, B. Schwab, D. Wehner, S. Albrecht, R. Ilg, D. Schüppel and T. v. Reden: *MAI Enviro - Vorstudie zur Lebenszyklusanalyse mit ökobilanzieller Bewertung relevanter Fertigungsprozessketten für CFK-Strukturen.* Fraunhofer-Verlag, Stuttgart, **2015**

[250] J. Huether, P. Schumann and K. A. Weidenmann: "Spent coffee ground as filler for fibre reinforced composites manufactured in a direct bulk moulding compound process." *Proceedings of ECCM 18 - 18th European conference on composite materials* 1–10, **2018**

[251] J. Huether, R. Maertens, O. Saburow, E. Seiler, T. Aicher and K. A. Weidenmann: "A holistic approach to use multi-scale fractions of dry carbon fibre production waste in filled bulk moulding compounds (BMC)." *Key Engineering Materials* Volume 742, 583–592, **2017**

[252] S. Zhandarov and E. Mäder: "Characterization of fiber/matrix interface strength: applicability of different tests, approaches and parameters." *Composites Science and Technology* Volume 65, 149–160, **2005**

[253] D. F. Adams and D. R. Doner: "Longitudinal shear loading of a unidirectional composite." *Journal of Composite Materials* Volume 1, 4–17, **1967**

[254] L. T. Drzal, N. Sugiura and D. Hook: "The role of chemical bonding and surface topography in adhesion between carbon fibers and epoxy matrices." *Composite Interfaces* Volume 4, 337–354, **1996**

[255] G. P. Tandon and N. J. Pagano: "Micromechanical analysis of the fiber push-out and re-push test." *Composites Science and Technology* Volume 58, 1709–1725, **1998**

[256] R. J. Kerans and T. A. Parthasarathy: "Theoretical analysis of the fiber pullout and pushout tests." *Journal of the American Ceramic Society* Volume 74, 1585–1596, **1991**

[257] L. S. Penn and E. R. Bowler: "A new approach to surface energy characterization for adhesive performance prediction." *Surface and Interface Analysis* Volume 3, 161–164, **1981**

[258] B. Miller, P. Muri and L. Rebenfeld: "A microbond method for determination of the shear strength of a fiber/resin interface." *Composites Science and Technology* Volume 28, 17–32, **1987**

[259] U. Gaur and B. Miller: "Microbond method for determination of the shear strength of a fiber/resin interface: evaluation of experimental parameters." *Composites Science and Technology* Volume 34, 35–51, **1989**

[260] D. B. Marshall: "An indentation method for measuring matrix-fiber frictional stresses in ceramic composites." *Journal of the American Ceramic Society* Volume 67, 259–260, **1984**

[261] B. Brylka, F. Fritzen, T. Böhlke and K. Weidenmann: "Study of experimental methods for interface problems based on virtual testing." *PAMM* Volume 10, 109–110, **2010**

[262] B. Brylka, F. Fritzen, T. Böhlke and K. A. Weidenmann: "Influence of micro-structure on fibre push-out tests." *PAMM* Volume 11, 141–142, **2011**

[263] S. Sockalingam and G. Nilakantan: "Fiber-matrix interface characterization through the microbond test." *International Journal of Aeronautical and Space Sciences* Volume 13, 282–295, **2012**

[264] S. Mohammadi-Jam and K. E. Waters: "Inverse gas chromatography applications: a review." *Advances in colloid and interface science* Volume 212, 21–44, **2014**

[265] A. van Asten, N. van Veenendaal and S. Koster: "Surface characterization of industrial fibers with inverse gas chromatography." *Journal of Chromatography A* Volume 888, 175–196, **2000**

[266] S. Brunauer, P. H. Emmett and E. Teller: "Adsorption of gases in multimolecular layers." *Journal of the American Chemical Society* Volume 60, 309–319, **1938**

[267] M. A. Nahil and P. T. Williams: "Recycling of carbon fibre reinforced polymeric waste for the production of activated carbon fibres." *Journal of Analytical and Applied Pyrolysis* Volume 91, 67–75, **2011**

[268] L. Lavielle and J. Schultz: "Surface properties of carbon fibers determined by inverse gas chromatography: role of pretreatment." *Langmuir : the ACS journal of surfaces and colloids* Volume 7, 978–98, **1991**

[269] J. Schultz, L. Lavielle and C. Martin: "The role of the interface in carbon fibre-epoxy composites." *The Journal of Adhesion* Volume 23, 45–60, **1987**

[270] S. Sheokand, S. R. Modi and A. K. Bansal: "Dynamic vapor sorption as a tool for characterization and quantification of amorphous content in predominantly crystalline materials." *Journal of pharmaceutical sciences* Volume 103, 3364–3376, **2014**

[271] P. York: "Solid-state properties of powders in the formulation and processing of solid dosage forms." *International Journal of Pharmaceutics* Volume 14, 1–28, **1983**

[272] S. Okubayashi, U. J. Griesser and T. Bechtold: "Moisture sorption/desorption behavior of various manmade cellulosic fibers." *Journal of Applied Polymer Science* Volume 97, 1621–1625, **2005**

[273] R. Oldenziel and H. Weber: "Introduction: Reconsidering recycling." *Contemporary European History* Volume 22, 347–370, **2013**

[274] R.-D. Weege: *Recyclinggerechtes Konstruieren.* VDI-Verl., Düsseldorf, **1981**

[275] The European Parliament and the council of the European Union: "Regulation No 510/2011 on setting emission performance standards for new light commercial vehicles as part of the Union's integrated approach to reduce CO2 emissions from light-duty vehicle." *Commission Statements* **2011**

[276] The European Parliament and the council of the European Union: "Council Directive 1999/31/EC on the landfill of waste." *Commission Statements* **1999**

[277] The European Parliament and the council of the European Union. "Directive 2000/53/EC on end-of life vehicles." *Commission Statements* **2000**

[278] E. Asmatulu, J. Twomey and M. Overcash: "Recycling of fiber-reinforced composites and direct structural composite recycling concept." *Journal of Composite Materials* Volume 48, 593–608, **2013**

[279] J. Rybicka, A. Tiwari and G. A. Leeke: "Technology readiness level assessment of composites recycling technologies." *Journal of Cleaner Production* Volume 112, 1001–1012, **2016**

[280] M. F. Ashby: "Software CES EduPack: Version 16.1.22.", **2016**

[281] G. Oliveux, L. O. Dandy and G. A. Leeke: "Current status of recycling of fibre reinforced polymers: review of technologies, reuse and resulting properties." *Progress in Materials Science* Volume 72, 61–99, **2015**

[282] F. Pottmeyer, M. Muth and K. A. Weidenmann: "Research of the Load Bearing Capacity of Shape-Optimized Metal Inserts Embedded in CFRP under Different Types of Stresses." *Key Engineering Materials* Volume 742, 636–643, **2017**

[283] P. Rupp, P. Elsner and K. A. Weidenmann: "Specific bending stiffness of in-mould-assembled hybrid sandwich structures with carbon fibre reinforced polymer face sheets and aluminium foam cores manufactured by a polyurethane-spraying process." *Journal of Sandwich Structures & Materials* Volume 21, 2779–2800, **2019**

[284] V. Sessner, M. Stoll, A. Feuvrier and K. A. Weidenmann: "Determination of the damping characteristics of fiber-metal-elastomer laminates using piezo-indicated-loss-factor experiments." *Key Engineering Materials* Volume 742, 325–332, **2017**

[285] Y. Yang, R. Boom, B. Irion, D.-J. van Heerden, P. Kuiper and H. de Wit: "Recycling of composite materials." *Chemical Engineering and Processing: Process Intensification* Volume 51, 53–68, **2012**

[286] S. Schweizer, A. Becker-Staines and T. Tröster: "Separation of hybrid structures for the reclaim of their single components." *Key Engineering Materials* Volume 742, 568–575, **2017**

[287] M. Blazsó: "Pyrolysis for recycling waste composites." In V. Goodship, editor, "Management, recycling and reuse of waste composites," Woodhead Publishing in materials, 102–121. Woodhead Pub, Oxford and Boca Raton, **2010**

[288] M. Y. Zakaria, A. B. Sulong, J. Sahari and H. Suherman: "Effect of the addition of milled carbon fiber as a secondary filler on the electrical conductivity of graphite/epoxy composites for electrical conductive material." *Composites Part B: Engineering* Volume 83, 75–80, **2015**

[289] M. Hengstermann, N. Raithel, A. Abdkader, M. M. Hasan and C. Cherif: "Development of new hybrid yarn construction from recycled carbon fibers for high performance composites. Part-I: basic processing of hybrid carbon fiber/polyamide 6 yarn spinning from virgin carbon fiber staple fibers." *Textile Research Journal* Volume 86, 1307–1317, **2016**

[290] J. Petterson and P. Nilsson: "Recycling of SMC and BMC in Standard Process Equipment." *Journal of Thermoplastic Composite Materials* Volume 7, 56–63, **1994**

[291] K. Ogi, T. Nishikawa, Y. Okano and I. Taketa: "Mechanical properties of ABS resin reinforced with recycled CFRP." *Advanced Composite Materials* Volume 16, 181–194, **2007**

[292] J. Takahashi, N. Matsutsuka, T. Okazumi, K. Uzawa, I. Ohsawa, K. Yamaguchi and A. Kitano: "Mechanical properties of recycled CFRP by injection molding method." *ICCM-16, Japan Society for Composite Materials, Kyoto, Japan* **2007**

[293] G. Schinner, J. Brandt and H. Richter: "Recycling carbon-fiber-reinforced thermoplastic composites." *Journal of Thermoplastic Composite Materials* Volume 9, 239–245, **1996**

[294] S. J. Pickering: "Recycling technologies for thermoset composite materials – current status." *Composites Part A: Applied Science and Manufacturing* Volume 37, 1206–1215, **2006**

[295] T. A. Turner, N. A. Warrior and S. J. Pickering: "Development of high value moulding compounds from recycled carbon fibres." *Plastics, Rubber and Composites* Volume 39, 151–156, **2010**

[296] B. C. Jin, X. Li, A. Jain, C. González, J. LLorca and S. Nutt: "Optimization of microstructures and mechanical properties of composite oriented strand board from reused prepreg." *Composite Structures* Volume 174, 389–398, **2017**

[297] G. Nilakantan and S. Nutt: "Reuse and upcycling of thermoset prepreg scrap: case study with out-of-autoclave carbon fiber/epoxy prepreg." *Journal of Composite Materials* Volume 52, 341–360, **2017**

[298] K. Ogi, T. Shinoda and M. Mizui: "Strength in concrete reinforced with recycled CFRP pieces." *Composites Part A: Applied Science and Manufacturing* Volume 36, 893–902, **2005**

[299] J. Palmer, L. Savage, O. R. Ghita and K. E. Evans: "Sheet moulding compound (SMC) from carbon fibre recyclate." *Composites Part A: Applied Science and Manufacturing* Volume 41, 1232–1237, **2010**

[300] K. Stoeffler, S. Andjelic, N. Legros, J. Roberge and S. B. Schougaard: "Polyphenylene sulfide (PPS) composites reinforced with recycled carbon fiber." *Composites Science and Technology* Volume 84, 65–71, **2013**

[301] M. L. Longana, N. Ong, H. Yu and K. D. Potter: "Multiple closed loop recycling of carbon fibre composites with the HiPerDiF (High Performance Discontinuous Fibre) method." *Composite Structures* Volume 153, 271–277, **2016**

[302] S. J. Pickering, Z. Liu, T. A. Turner and K. H. Wong: "Applications for carbon fibre recovered from composites." *IOP Conference Series: Materials Science and Engineering* Volume 139, 1–18, **2016**

[303] S. J. Pickering, R. M. Kelly, J. R. Kennerley, C. D. Rudd and N. J. Fenwick: "A fluidised-bed process for the recovery of glass fibres from scrap thermoset composites." *Composites Science and Technology* Volume 60, 509–523, **2000**

[304] J. L. Thomason, L. Yang and R. Meier: "The properties of glass fibres after conditioning at composite recycling temperatures." *Composites Part A: Applied Science and Manufacturing* Volume 61, 201–208, **2014**

[305] S. Sakka: "Effects of reheating on strength of glass fibers." *Bulletin of the Institute for Chemical Research, Kyoto University* 316–320, **1956**

[306] S. R. Naqvi, H. M. Prabhakara, E. A. Bramer, W. Dierkes, R. Akkerman and G. Brem: "A critical review on recycling of end-of-life carbon fibre/glass fibre reinforced composites waste using pyrolysis towards a circular economy." *Resources, Conservation and Recycling* Volume 136, 118–129, **2018**

[307] R. Piñero-Hernanz, C. Dodds, J. Hyde, J. García-Serna, M. Poliakoff, E. Lester, M. J. Cocero, S. Kingman, S. Pickering and K. H. Wong: "Chemical recycling of carbon fibre reinforced composites in nearcritical and supercritical water." *Composites Part A: Applied Science and Manufacturing* Volume 39, 454–461, **2008**

[308] R. Piñero-Hernanz, J. García-Serna, C. Dodds, J. Hyde, M. Poliakoff, M. J. Cocero, S. Kingman, S. Pickering and E. Lester: "Chemical recycling of carbon fibre composites using alcohols under subcritical and supercritical conditions." *The Journal of Supercritical Fluids* Volume 46, 83–92, **2008**

[309] J. Jiang, G. Deng, X. Chen, X. Gao, Q. Guo, C. Xu and L. Zhou: "On the successful chemical recycling of carbon fiber/epoxy resin composites under the mild condition." *Composites Science and Technology* Volume 151, 243–251, **2017**

[310] Y. Liu, L. Meng, Y. Huang and J. Du: "Recycling of carbon/epoxy composites." *Journal of Applied Polymer Science* Volume 94, 1912–1916, **2004**

[311] A. Fernández, C. S. Lopes, C. González and F. A. López: "Characterization of carbon fibers recovered by pyrolysis of cured prepregs and their reuse in new composites." In R. Khanna and R. Cayumil, editors, "Recent Developments in the Field of Carbon Fibers," InTech, London, **2018**

[312] A. Greco, A. Maffezzoli, G. Buccoliero, F. Caretto and G. Cornacchia: "Thermal and chemical treatments of recycled carbon fibres for improved adhesion to polymeric matrix." *Journal of Composite Materials* Volume 47, 369–377, **2013**

[313] R. A. Witik, R. Teuscher, V. Michaud, C. Ludwig and J.-A. E. Månson: "Carbon fibre reinforced composite waste: An environmental assessment of recycling, energy recovery and landfilling." *Composites Part A: Applied Science and Manufacturing* Volume 49, 89–99, **2013**

[314] Y. Zhou, Y. Wang, Y. Xia and S. Jeelani: "Tensile behavior of carbon fiber bundles at different strain rates." *Materials Letters* Volume 64, 246–248, **2010**

[315] N. Dilsiz and J. Wightman: "Surface analysis of unsized and sized carbon fibers." *Carbon* Volume 37, 1105–1114, **1999**

[316] Y. Yang, C. Lu, X. Su and X. Wang: "Effects of emulsion sizing with nano-SiO2 on interfacial properties of carbon fibers/epoxy composites." *Journal of Materials Science* Volume 42, 6347–6352, **2007**

[317] B. Fernández, A. Arbelaiz, A. Valea, F. Mujika and I. Mondragon: "A comparative study on the influence of epoxy sizings on the mechanical performance of woven carbon fiber-epoxy composites." *Polymer Composites* Volume 25, 319–330, **2004**

[318] D. Bücheler, A. Kaiser and F. Henning: "Using thermogravimetric analysis to determine carbon fiber weight percentage of fiber-reinforced plastics." *Composites Part B: Engineering* Volume 106, 218–223, **2016**

[319] C. Kunzmann, J. Moosburger-Will and S. Horn: "High-resolution imaging of the nanostructured surface of polyacrylonitrile-based fibers." *Journal of Materials Science* Volume 51, 9638–9648, **2016**

[320] J. Jäger, J. Moosburger-Will and S. Horn: "Determination of nanoroughness of carbon fibers by atomic force microscopy." *Journal of Materials Science* Volume 48, 6803–6810, **2013**

[321] Deutsches Institut für Normung: "DIN EN 843-5: Hochleistungskeramik - Mechanische Eigenschaften monolithischer Keramik bei Raumtemperatur - Teil 5: Statistische Auswertung." **2006**

[322] E. Mäder, C. Scheffler, S. Fliescher, U. Mörschel, C. Poitzsch and A. Miene: "Fimatest – a new testing system to determine the fibre-matrix adhesion strength by means of pull-out tests." *Proceedings of 56th Dornbirn-MFC* Volume 56, **2017**

[323] D02 Committee: *ASTM standard D2638–10: Test Method for Real Density of Calcined Petroleum Coke by Helium Pycnometer*. ASTM International, West Conshohocken, PA, **2015**

[324] T. J. Rude, L. H. Strait (Jr.) and L. Ruhala: "Measurement of fiber density by helium pycnometry." *Journal of Composite Materials* Volume 34, 1948–1958, **2000**

[325] K. P. Menard: *Dynamic Mechanical Analysis: A Practical Introduction*. Taylor & Francis, Hoboken, 2nd edition, **2008**

[326] D20 Committee: *ASTM standard D5023–15: Test Method for Plastics: Dynamic Mechanical Properties: In Flexure (Three-Point Bending)*. ASTM International, West Conshohocken, PA, **2015**

[327] E37 Committee: *ASTM standard E1640–18: Test Method for Assignment of the Glass Transition Temperature By Dynamic Mechanical Analysis*. ASTM International, West Conshohocken, PA, **2018**

[328] Deutsches Institut für Normung: "DIN EN ISO 527-1: Kunststoffe - Bestimmung der Zugeigenschaften. Teil 1: Allgemeine Grundsätze." **1996**

[329] Deutsches Institut für Normung: "DIN EN ISO 527-4: Kunststoffe - Bestimmung der Zugeigenschaften. Teil 4: Prüfbedingungen für isotrop und anisotrop faserverstärkte Kunststoffverbundwerkstoffe." **1997**

[330] B. Shi, Y. Wang and L. Jia: "Comparison of Dorris-Gray and Schultz methods for the calculation of surface dispersive free energy by inverse gas chromatography." *Journal of chromatography. A* Volume 1218, 860–862, **2011**

[331] Y. Liu and S. Kumar: "Recent progress in fabrication, structure, and properties of carbon fibers." *Polymer Reviews* Volume 52, 234–258, **2012**

[332] G. Jiang, S. J. Pickering, E. H. Lester, T. Turner, K. Wong and N. Warrior: "Characterisation of carbon fibres recycled from carbon fibre/epoxy resin composites using supercritical n-propanol." *Composites Science and Technology* Volume 69, 192–198, **2009**

[333] D. J. Johnson: "Structure-property relationships in carbon fibres." *Journal of Physics D: Applied Physics* Volume 20, 286–291, **1987**

[334] B. A. Newcomb, L. A. Giannuzzi, K. M. Lyons, P. V. Gulgunje, K. Gupta, Y. Liu, M. Kamath, K. McDonald, J. Moon, B. Feng, G. P. Peterson, H. G. Chae and S. Kumar: "High resolution transmission electron microscopy study on polyacrylonitrile/carbon nanotube based carbon fibers and the effect of structure development on the thermal and electrical conductivities." *Carbon* Volume 93, 502–514, **2015**

[335] L. Fischer and W. Ruland: "The influence of graphitization on the mechanical properties of carbon fibers." *Colloid and Polymer Science* Volume 258, 917–922, **1980**

[336] R. B. Mathur, O. P. Bahl, V. K. Matta and K. C. Nagpal: "Modification of PAN precursor—Its influence on the reaction kinetics." *Carbon* Volume 26, 295–301, **1988**

[337] L. O. Meyer, K. Schulte and E. Grove-Nielsen: "CFRP-recycling following a pyrolysis route: process optimization and potentials." *Journal of Composite Materials* Volume 43, 1121–1132, **2009**

[338] M. H. Akonda, C. A. Lawrence and B. M. Weager: "Recycled carbon fibre-reinforced polypropylene thermoplastic composites." *Composites Part A: Applied Science and Manufacturing* Volume 43, 79–86, **2011**

[339] E. Lester, S. Kingman, K. H. Wong, C. Rudd, S. Pickering and N. Hilal: "Microwave heating as a means for carbon fibre recovery from polymer composites: A technical feasibility study." *Materials Research Bulletin* Volume 39, 1549–1556, **2004**

[340] H. L. H. Yip, S. J. Pickering and C. D. Rudd: "Characterisation of carbon fibres recycled from scrap composites using fluidised bed process." *Plastics, Rubber and Composites* Volume 31, 278–282, **2002**

[341] J. R. Hyde, E. Lester, S. Kingman, S. Pickering and K. H. Wong: "Supercritical propanol, a possible route to composite carbon fibre recovery: a viability study." *Composites Part A: Applied Science and Manufacturing* Volume 37, 2171–2175, **2006**

[342] A. Kafi, M. Huson, C. Creighton, J. Khoo, L. Mazzola, T. Gengenbach, F. Jones and B. Fox: "Effect of surface functionality of pan-based carbon fibres on the mechanical performance of carbon/epoxy composites." *Composites Science and Technology* Volume 94, 89–95, **2014**

[343] K. K. Chee Ho, H. Qian and A. Bismarck: "Carbon fiber: surface properties." *Wiley encyclopedia of composites* Volume 13, 1–11, **2011**

[344] E. Pamula and P. G. Rouxhet: "Bulk and surface chemical functionalities of type III PAN-based carbon fibres." *Carbon* Volume 41, 1905–1915, **2003**

[345] Q. Li, J. S. Church, M. Naebe and B. L. Fox: "Interfacial characterization and reinforcing mechanism of novel carbon nanotube – carbon fibre hybrid composites." *Carbon* Volume 109, 74–86, **2016**

[346] A. Kafi, Q. Li, T. Chaffraix, J. Khoo, T. Gengenbach and K. J.-C. Magniez: "Surface treatment of carbon fibres for interfacial property enhancement in composites via surface deposition of water soluble POSS nanowhiskers." *Polymer* Volume 137, 97–106, **2018**

[347] Kafi, Abdullah and Henderson, Luke and Servinis, Linden and Huson, Mickey and Gengenbach, Thomas and Khoo, Jiyi and Garschke, Claudia and Fox, Bronwyn: "New surface treatments for low cost carbon fiber." *SAMPE 2013: Proceedings of the Society for the Advancement of Material and Process Engineering* 456–469, **2013**

[348] H. N. Petersen, Y. Kusano, P. Brøndsted and K. Almdal: "The influcnce of removing sizing on strength and stiffness of conventional and high modulus E-glass fibres." *IOP Conference Series: Materials Science and Engineering* Volume 139, 1–6, **2016**

[349] G. Lamanna and A. Ceparano: "Mechanical characterization of sheet moulding composites for the automotive industry." *The Open Materials Science Journal* Volume 8, 108–113, **2014**

[350] A. Trauth, P. Pinter and K. Weidenmann: "Investigation of quasi-static and dynamic material properties of a structural sheet molding compound combined with acoustic emission damage analysis." *Journal of Composites Science* Volume 1, 1–18, **2017**

[351] E. S. Greenhalgh: *Failure analysis and fractography of polymer composites*. Woodhead Publishing in materials. Woodhead Pub, Cambridge, **2009**

[352] P. Feraboli, H. Kawakami, B. Wade, F. Gasco, L. DeOto and A. Masini: "Recyclability and reutilization of carbon fiber fabric/epoxy composites." *Journal of Composite Materials* Volume 46, 1459–1473, **2012**

[353] D. Rosato: "Advances in moulding reinforced plastic parts for automotive applications. A review of the industry in the USA." *Composites* Volume 4, 208–217, **1973**

[354] J. N. Epel: "Comparison of low profile premix and sheet molding compound applications." In "SAE Technical Paper Series," SAE Technical Paper Series. SAE International, Warrendale, PA, United States, **1971**

List of Figures

List of Tables

Schriftenreihe des Instituts für Angewandte Materialien

ISSN 2192-9963

Band 1 Prachai Norajitra
Divertor Development for a Future Fusion Power Plant. 2011
ISBN 978-3-86644-738-7

Band 2 Jürgen Prokop
Entwicklung von Spritzgießsonderverfahren zur Herstellung von Mikrobauteilen durch galvanische Replikation. 2011
ISBN 978-3-86644-755-4

Band 3 Theo Fett
New contributions to R-curves and bridging stresses – Applications of weight functions. 2012
ISBN 978-3-86644-836-0

Band 4 Jérôme Acker
Einfluss des Alkali/Niob-Verhältnisses und der Kupferdotierung auf das Sinterverhalten, die Strukturbildung und die Mikrostruktur von bleifreier Piezokeramik $(K_{0,5}Na_{0,5})NbO_3$. 2012
ISBN 978-3-86644-867-4

Band 5 Holger Schwaab
Nichtlineare Modellierung von Ferroelektrika unter Berücksichtigung der elektrischen Leitfähigkeit. 2012
ISBN 978-3-86644-869-8

Band 6 Christian Dethloff
Modeling of Helium Bubble Nucleation and Growth in Neutron Irradiated RAFM Steels. 2012
ISBN 978-3-86644-901-5

Band 7 Jens Reiser
Duktilisierung von Wolfram. Synthese, Analyse und Charakterisierung von Wolframlaminaten aus Wolframfolie. 2012
ISBN 978-3-86644-902-2

Band 8 Andreas Sedlmayr
Experimental Investigations of Deformation Pathways in Nanowires. 2012
ISBN 978-3-86644-905-3

Die Bände sind unter www.ksp.kit.edu als PDF frei verfügbar oder als Druckausgabe bestellbar.

Band 9 Matthias Friedrich Funk
Microstructural stability of nanostructured fcc metals during cyclic deformation and fatigue. 2012
ISBN 978-3-86644-918-3

Band 10 Maximilian Schwenk
Entwicklung und Validierung eines numerischen Simulationsmodells zur Beschreibung der induktiven Ein- und Zweifrequenzrandschichthärtung am Beispiel von vergütetem 42CrMo4. 2012
ISBN 978-3-86644-929-9

Band 11 Matthias Merzkirch
Verformungs- und Schädigungsverhalten der verbundstranggepressten, federstahldrahtverstärkten Aluminiumlegierung EN AW-6082. 2012
ISBN 978-3-86644-933-6

Band 12 Thilo Hammers
Wärmebehandlung und Recken von verbundstranggepressten Luftfahrtprofilen. 2013
ISBN 978-3-86644-947-3

Band 13 Jochen Lohmiller
Investigation of deformation mechanisms in nanocrystalline metals and alloys by in situ synchrotron X-ray diffraction. 2013
ISBN 978-3-86644-962-6

Band 14 Simone Schreijäg
Microstructure and Mechanical Behavior of Deep Drawing DC04 Steel at Different Length Scales. 2013
ISBN 978-3-86644-967-1

Band 15 Zhiming Chen
Modelling the plastic deformation of iron. 2013
ISBN 978-3-86644-968-8

Band 16 Abdullah Fatih Çetinel
Oberflächendefektausheilung und Festigkeitssteigerung von niederdruckspritzgegossenen Mikrobiegebalken aus Zirkoniumdioxid. 2013
ISBN 978-3-86644-976-3

Band 17 Thomas Weber
Entwicklung und Optimierung von gradierten Wolfram/ EUROFER97-Verbindungen für Divertorkomponenten. 2013
ISBN 978-3-86644-993-0

Band 18 Melanie Senn
Optimale Prozessführung mit merkmalsbasierter Zustandsverfolgung. 2013
ISBN 978-3-7315-0004-9

Band 19 Christian Mennerich
Phase-field modeling of multi-domain evolution in ferromagnetic shape memory alloys and of polycrystalline thin film growth. 2013
ISBN 978-3-7315-0009-4

Band 20 Spyridon Korres
On-Line Topographic Measurements of Lubricated Metallic Sliding Surfaces. 2013
ISBN 978-3-7315-0017-9

Band 21 Abhik Narayan Choudhury
Quantitative phase-field model for phase transformations in multi-component alloys. 2013
ISBN 978-3-7315-0020-9

Band 22 Oliver Ulrich
Isothermes und thermisch-mechanisches Ermüdungsverhalten von Verbundwerkstoffen mit Durchdringungsgefüge (Preform-MMCs). 2013
ISBN 978-3-7315-0024-7

Band 23 Sofie Burger
High Cycle Fatigue of Al and Cu Thin Films by a Novel High-Throughput Method. 2013
ISBN 978-3-7315-0025-4

Band 24 Michael Teutsch
Entwicklung von elektrochemisch abgeschiedenem LIGA-Ni-Al für Hochtemperatur-MEMS-Anwendungen. 2013
ISBN 978-3-7315-0026-1

Band 25 Wolfgang Rheinheimer
Zur Grenzflächenanisotropie von $SrTiO_3$. 2013
ISBN 978-3-7315-0027-8

Band 26 Ying Chen
Deformation Behavior of Thin Metallic Wires under Tensile and Torsional Loadings. 2013
ISBN 978-3-7315-0049-0

Die Bände sind unter www.ksp.kit.edu als PDF frei verfügbar oder als Druckausgabe bestellbar.

Band 27 Sascha Haller
Gestaltfindung: Untersuchungen zur Kraftkegelmethode. 2013
ISBN 978-3-7315-0050-6

Band 28 Nicht erschienen

Band 29 Gunnar Picht
Einfluss der Korngröße auf ferroelektrische Eigenschaften dotierter $Pb(Zr_{1-x}Ti_x)O_3$ Materialien. 2013
ISBN 978-3-7315-0106-0

Band 30 Esther Held
Eigenspannungsanalyse an Schichtverbunden mittels inkrementeller Bohrlochmethode. 2013
ISBN 978-3-7315-0127-5

Band 31 Pei He
On the structure-property correlation and the evolution of Nanofeatures in 12-13.5% Cr oxide dispersion strengthened ferritic steels. 2014
ISBN 978-3-7315-0141-1

Band 32 Jan Hoffmann
Ferritische ODS-Stähle – Herstellung, Umformung und Strukturanalyse. 2014
ISBN 978-3-7315-0157-2

Band 33 Wiebke Sittel
Entwicklung und Optimierung des Diffusionsschweißens von ODS Legierungen. 2014
ISBN 978-3-7315-0182-4

Band 34 Osama Khalil
Isothermes Kurzzeitermüdungsverhalten der hoch-warmfesten Aluminium-Knetlegierung 2618A (AlCu2Mg1,5Ni). 2014
ISBN 978-3-7315-0208-1

Band 35 Nicht erschienen

Band 36 Christoph Hage
Grundlegende Aspekte des 2K-Metallpulverspritzgießens. 2014
ISBN 978-3-7315-0217-3

Band 37 Bartłomiej Albiński
Instrumentierte Eindringprüfung bei Hochtemperatur für die Charakterisierung bestrahlter Materialien. 2014
ISBN 978-3-7315-0221-0

Band 38 Tim Feser
Untersuchungen zum Einlaufverhalten binärer alpha-Messinglegierungen unter Ölschmierung in Abhängigkeit des Zinkgehaltes. 2014
ISBN 978-3-7315-0224-1

Band 39 Jörg Ettrich
Fluid Flow and Heat Transfer in Cellular Solids. 2014
ISBN 978-3-7315-0241-8

Band 40 Melanie Syha
Microstructure evolution in strontium titanate Investigated by means of grain growth simulations and x-ray diffraction contrast tomography experiments. 2014
ISBN 978-3-7315-0242-5

Band 41 Thomas Haas
Mechanische Zuverlässigkeit von gedruckten und gasförmig abgeschiedenen Schichten auf flexiblem Substrat. 2014
ISBN 978-3-7315-0250-0

Band 42 Aron Kneer
Numerische Untersuchung des Wärmeübertragungsverhaltens in unterschiedlichen porösen Medien. 2014
ISBN 978-3-7315-0252-4

Band 43 Manuel Feuchter
Investigations on Joule heating applications by multiphysical continuum simulations in nanoscale systems. 2014
ISBN 978-3-7315-0261-6

Band 44 Alexander Vondrous
Grain growth behavior and efficient large scale simulations of recrystallization with the phase-field method. 2014
ISBN 978-3-7315-0280-7

Band 45 Tobias Kennerknecht
Fatigue of Micro Molded Materials – Aluminum Bronze and Yttria Stabilized Zirconia. 2014
ISBN 978-3-7315-0293-7

Band 46 Christopher Scherr
Elektrochemisches Verhalten von Lithium-Schwefel-Zellen mit unterschiedlicher Kathodenstruktur. 2015
ISBN 978-3-7315-0296-8

Band 47 Konstantin Frölich
Der Decal-Prozess zur Herstellung katalysatorbeschichteter Membranen für PEM-Brennstoffzellen. 2015
ISBN 978-3-7315-0334-7

Band 48 Benedikt Haspel
Werkstoffanalytische Betrachtung der Eigenschaften von mittels neuartiger RTM-Fertigungsprozesse hergestellten glasfaserverstärkten Polymerverbunden. 2015
ISBN 978-3-7315-0337-8

Band 49 Marco Berghoff
Skalenübergreifende Modellierung und Optimierung vom atomistischen kristallinen Phasenfeldmodell bis zur mesoskopischen Phasenfeldmethode. 2015
ISBN 978-3-7315-0416-0

Band 50 Michael Selzer
Mechanische und Strömungsmechanische Topologieoptimierung mit der Phasenfeldmethode. 2016
ISBN 978-3-7315-0431-3

Band 51 Michael Mahler
Entwicklung einer Auswertemethode für bruchmechanische Versuche an kleinen Proben auf der Basis eines Kohäsivzonenmodells. 2016
ISBN 978-3-7315-0441-2

Band 52 Christoph Bohnert
Numerische Untersuchung des Verformungs- und Bruchverhaltens von einkristallinem Wolfram auf mikroskopischer Ebene. 2016
ISBN 978-3-7315-0444-3

Band 53 Stefan Guth
Schädigung und Lebensdauer von Nickelbasislegierungen unter thermisch-mechanischer Ermüdungsbeanspruchung bei verschiedenen Phasenlagen. 2016
ISBN 978-3-7315-0445-0

Band 54 Markus Klinsmann
The Effects of Internal Stress and Lithium Transport on Fracture in Storage Materials in Lithium-Ion Batteries. 2016
ISBN 978-3-7315-0455-9

Band 55 Thomas Straub
Experimental Investigation of Crack Initiation in Face-Centered Cubic Materials in the High and Very High Cycle Fatigue Regime. 2016
ISBN 978-3-7315-0471-9

Band 56 Maren Lepple
Kupfer- und Eisenoxide als Konversions-Elektrodenmaterialien für Lithium-Ionen-Batterien: Thermodynamische und Elektrochemische Untersuchungen. 2016
ISBN 978-3-7315-0482-5

Band 57 Stefan Andreas Slaby
Charakterisierung und Bewertung der Zug- und Ermüdungseigenschaften von Mikrobauteilen aus 17-4PH Edelstahl. Ein Vergleich von mikropulverspritzgegossenem und konventionell hergestelltem Material. 2017
ISBN 978-3-7315-0484-9

Band 58 Kumar Ankit
Phase-field modeling of microstructural pattern formation in alloys and geological veins. 2016
ISBN 978-3-7315-0491-7

Band 59 Kuo Zhang
Characterization and Modeling of the Ratcheting Behavior of the Ferritic-Martensitic Steel P91. 2017
ISBN 978-3-7315-0503-7

Band 60 Nicht erschienen

Band 61 Fabian Lemke
Untersuchung des Sinterverhaltens von $SrTiO_3$ unter Berücksichtigung der Defektchemie. 2016
ISBN 978-3-7315-0510-5

Band 62 Johannes Kümmel
Detaillierte Analyse der Aufbauschneidenbildung bei der Trockenzerspanung von Stahl C45E mit Berücksichtigung des Werkzeugverschleißes. 2016
ISBN 978-3-7315-0518-1

Band 63 László Hagymási
Modellierung der Stoffübertragung beim Niederdruckcarbonitrieren mit Ammoniak und Acetylen. 2016
ISBN 978-3-7315-0568-6

Band 64 Reza Eslami
A novel micro-mechanical model for prediction of multiaxial high cycle fatigue at small scales. 2017
ISBN 978-3-7315-0583-9

Band 65 Sebastian Schulz
Phase-field simulations of multi-component solidification and coarsening based on thermodynamic datasets. 2017
ISBN 978-3-7315-0618-8

Band 66 Markus Stricker
Die Übertragung von mikrostrukturellen Eigenschaften aus der diskreten Versetzungsdynamik in Kontinuumsbeschreibungen. 2017
ISBN 978-3-7315-0658-4

Band 67 Luis Straßberger
Untersuchung und Modellierung des viskoplastischen Verformungsverhaltens oxidpartikelverstärkter Stähle. 2018
ISBN 978-3-7315-0674-4

Band 68 Mark Wobrock
Microplasticity of idealized single crystalline Ag cantilevers characterized with methods of high resolution. 2017
ISBN 978-3-7315-0682-9

Band 69 Amritesh Kumar
Micromechanical study on the deformation behaviour of directionally solidified NiAl-Cr eutectic composites. 2017
ISBN 978-3-7315-0694-2

Band 70 Johannes Hötzer
Massiv-parallele und großskalige Phasenfeldsimulationen zur Untersuchung der Mikrostrukturentwicklung. 2017
ISBN 978-3-7315-0693-5

Band 71 Thomas Hupfer
Herstellung von LATP für den Einsatz als Festkörperelektrolyt und dessen Eigenschaften. 2017
ISBN 978-3-7315-0702-4

Band 72 Florentin Pottmeyer
Schädigungsverhalten von in CFK-Laminaten eingebetteten Inserts unter bauteilnahen Beanspruchungen. 2017
ISBN 978-3-7315-0719-2

Band 73 Andres Höweling
Untersuchung der Hochvoltstabilität und Tiefentladung von dotierten $LiNi_{0,5}Mn_{1,5}O_4$-Hochvoltspinellen. 2018
ISBN 978-3-7315-0728-4

Band 74 Tabea Gisela Schwark
Deformation and Fracture Properties of the Soft Magnetic Composite Somaloy 700 3P on Different Length Scales. 2018
ISBN 978-3-7315-0759-8

Band 75 Klaudia Lichtenberg
Metallmatrixverbunde mit Verstärkungselementen aus metallischem Glas $Ni_{60}Nb_{20}Ta_{20}$ – Herstellung und Charakterisierung. 2018
ISBN 978-3-7315-0782-6

Band 76 Claudio Findeisen
Charakterisierung und Modellierung von instabilen Metamaterialien. 2019
ISBN 978-3-7315-0869-4

Band 77 Nilesha Mishra
Influence of strain on the functionality of ink-jet printed thin films and devices on flexible substrates. 2019
ISBN 978-3-7315-0853-3

Band 78 Simon Werner Bonk
Plastische Verformungsmechanismen in hochgradig kaltgewalzten, ultrafeinkörnigen Wolframblechen. 2019
ISBN 978-3-7315-0878-6

Band 79 Tim Gräning
Herstellung, Charakterisierung und Optimierung von austenitischen ODS Stählen. 2019
ISBN 978-3-7315-0732-1

Band 80 Peter Rupp
Herstellung, Prüfung und Modellierung neuartiger hybrider Aluminiumschaum-CFK-Sandwichverbunde. 2019
ISBN 978-3-7315-0880-9

Band 81 Benjamin Sebastian Ehreiser
Einfluss mechanischer Lasten auf die Herstellung von Stahl-Glaskeramik-Verbunden. 2019
ISBN 978-3-7315-0954-7

Band 82 Hans Giel
Weiterentwicklung experimenteller Methoden zur Ermittlung thermodynamischer Werkstoffdaten von Lithium-Ionen-Batterien. 2020
ISBN 978-3-7315-0981-3

Band 83 Anna Trauth
Characterisation and Modelling of Continuous-Discontinuous Sheet Moulding Compound Composites for Structural Applications. 2020
ISBN 978-3-7315-0950-9

Band 84 Jonas Johannes Hüther
The Impact of Recycling on the Fibre and the Composite Properties of Carbon Fibre Reinforced Plastics. 2020
ISBN 978-3-7315-0983-7